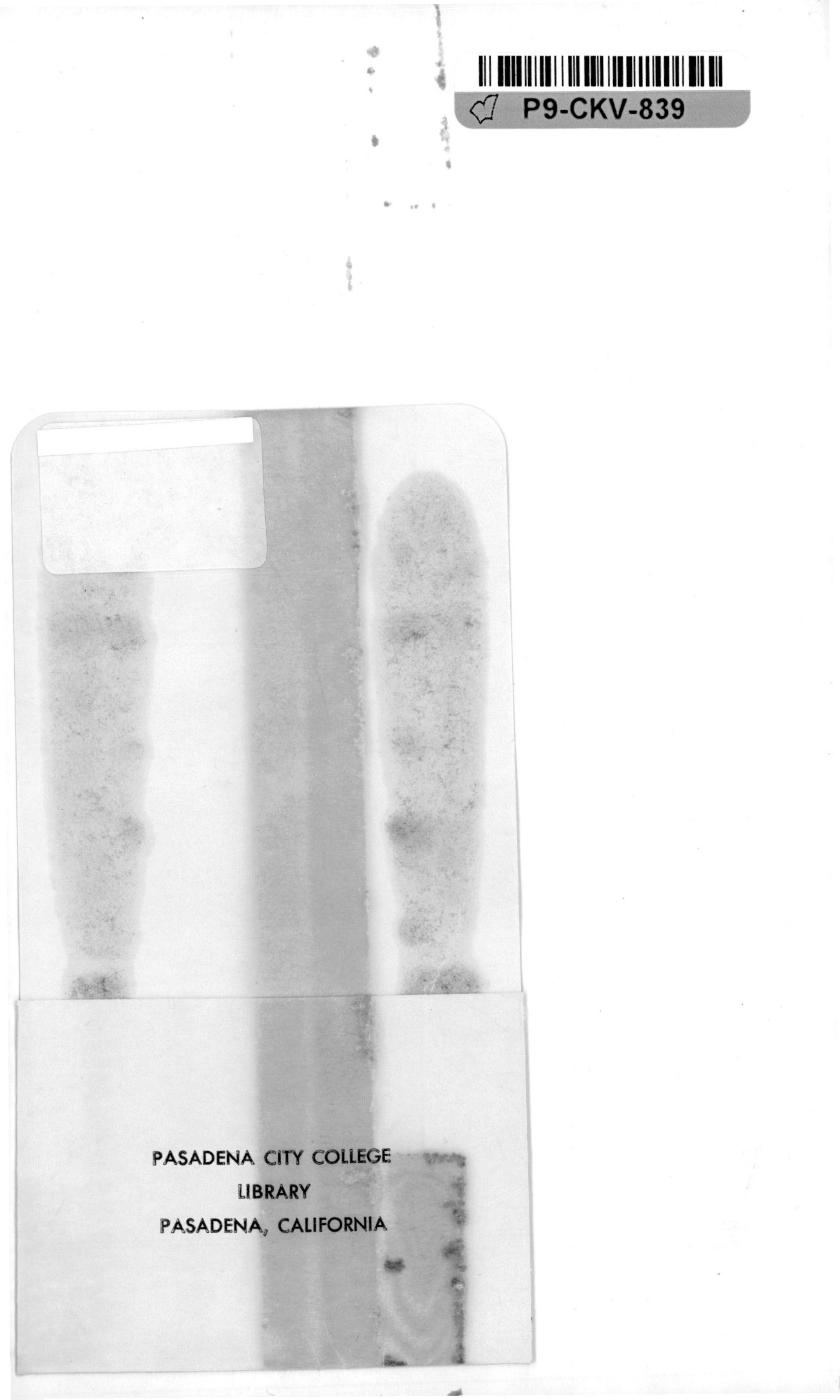

P9-CKV-839
PASADENA CITY COLLEGE
LIBRARY
PASADENA, CALIFORNIA

Listening and Voice

LISTENING AND VOICE

A Phenomenology of Sound

DON IHDE

Ohio University Press
Athens, Ohio

Copyright © 1976 by Don Ihde
ISBN 0-8214-0201-3
Library of Congress Catalog Number LC 76-8302
Printed in the United States of America by Oberlin Printing Co., Inc.

All rights reserved.

For

Leslie, Lisa, and Eric

WITHDRAWN

772315

CONTENTS

PREFACE

I have attempted to put this investigation in as straightforward and simple a fashion as possible. For this reason, although I have undertaken extensive studies in related fields such as the physiology of hearing, acoustics, and musical theory, references to these are implicit.

Except in the introductory remarks the same implicitness remains the case with the giants of the phenomenological tradition. I have forsworn any lengthy discussions of Husserl, Heidegger, Merleau-Ponty, Ricoeur, or Sartre; although all may be seen by the perceptive reader to be lurking beneath the surface.

Moreover, I have also chosen to describe things autobiographically to make the narrative even more straightforward. Because of this I run the risk that the book may be taken as merely autobiographical when its presumption is at least a bit more pretentious. It is intended as a prolegomena to an ontology of listening with suggestions for the implications of a philosophy of sound. And, in fact, while the style of the illustrations here is autobiographical, the investigations themselves took place over seven years and involved classroom investigation and much intersubjective research. In some instances studies in empirical psychology also suggested inquiry in a more phenomenological vein. Thus the danger of taking the studies as being mere assertions should be avoided. They should be taken as exercises in the application of a stricter phenomenological use of *variations* which have in every instance been cross-checked with other persons' experiences.

It has been my hope that my adoption of this style may gain more than it loses by giving a sense of *doing* a phenomenological investigation and by dealing with it in a language which I have attempted to make as clear as my abilities allow. In the process I may well lose the attention of some who prefer extensive proof texting and multiple footnotes to show the indebtedness of which I am only too painfully aware. But I also wish to gain what I hope will be a sense of the excitement which can come from getting under way in initiating a phenomenology of listening and voice.

There is one other preliminary problem I wish to point out. To do a phenomenology of sound in a *book* is itself something of a functional "contradiction." A book is read and its words are seen rather than heard. There are vast differences between hearing voices and reading words; yet the distance between the language embodied visually and that which is heard is sometimes broached. Sometimes there is a "singing" of voice *in* writing. I have often been shocked at "hearing" a friend's voice upon reading his or her latest article or book. The other sounds through in an auditory adherence to what is ordinarily soundless. The same phenomenon occurs with trained musicians who can "hear" the music they see when reading a score.

These phenomena themselves are not only perplexing but intriguing and must be part of the sense of the investigation. Ultimately, involvement with the world must show itself as well. That will be part of the task of the text and part of the terrain of the investigation itself.

Finally, those sources of explicit and concrete help which made the book possible should be acknowledged. Support for the research came in the forms of Summer Faculty Fellowships awarded by the State University of New York and a Senior Fellowship awarded by the National Endowment for the Humanities in 1972. I wish to thank the many preliminary readers of the manuscript for their suggestions and criticisms. In particular I am grateful to Professors David Carr, Edward Casey, and John O'Neill; and to David Allison, Roger Bell, and Elyse Glass at State University of New York at Stony Brook. A special note of thanks should go to my wife, Carolyn, for typing and correcting the manuscript and to my assistant, Robin Elliser, for her work correcting copy.

PART ONE

INTRODUCTION

CHAPTER ONE

IN PRAISE OF SOUND

The beginning of man is in the midst of *word*. And the center of word is in breath and sound, in listening and speaking. In the ancient mythologies the word for soul was often related to the word for breath. In the biblical myth of the creation, God breathes life into Adam, and that breath is both life and word.

Today mythical thought is still repeated in other ways. We know that we live immersed in a vast but invisible ocean of air which surrounds us and permeates us and without which our life must necessarily escape us. For even when we humans wander far from the surface of the earth to that of the moon or deep into the sea, we must take with us packaged envelopes of air which we inhale and exhale. But in the words about breath there lurk ancient significances by which we take in the haleness or health of the air which for the ancients was spirit. From breath and the submersion in air also comes *in-spire*, "to take in spirit," and upon a final *ex-halation* we ex-(s)pire, and the spirit leaves us without life. Thus still with us, hidden in our language, is something of the ontology of Anaximenes who, concerning the air, thought, "As our souls, being air, hold us together, so breath and air embrace the entire universe."[1]

But the air which is breathed is not neutral or lifeless, for it has its life in *sound* and *voice*. Its sound ranges from the barely or not-at-all noticed background of our own breathing to the noises of the world and the singing of word and song among humans. The silence of the invisible comes to life in sound. For the human listener there is a multiplicity of senses in which *there is word in the wind*.

From a thoroughly contemporary source the importance of soundful significance may be discerned today as well. This new interest arises from various fronts of the contemporary sciences and philosophies. In philosophy there can be no doubt that questions of language and speech have been of great if not dominant importance in current philosophy. If on the one side that interest has been primarily in logic and syntax, as is the case with the Anglo-American philosophies, and on the other the interest has been the birth of meaning in speech in Continental thought, the question of word has been a central concern of the twentieth century. There has also arisen and flourished a whole series of linguistic sciences which relate to the question of word: phonetics, semiology, structural and generative linguistics, and the diverse schools of semantics.

Yet after the critical thinker has studied and read through these disciplines with their admittedly brilliant advances, there can remain a doubt that everything essential has been noted. For there appears in the very proliferation of disciplines addressed to the question of word a division which leaves word disincarnate. On the one side are the disciplines which address the structure, the form, the mechanics of language. Its surface and depth rules which produce significances are conceived of almost without the sense of enactment by a speaker in what may be termed a mechanics of language. The philosopher, concerned with comprehensiveness, must eventually call for attention to the *word as soundful.* On the other side the sciences which attend to the soundful, from phonetics to acoustics, do so as if the sound were bare and empty of significance in a physics of the soundful. And the philosopher, concerned with the roots of reflection in human experience, must eventually also listen to the *sounds as meaningful.*

There is a third source of the contemporary interest in sound and listening which, while so familiar as to be taken for granted, includes within it a subtle and profound transformation of experience itself as our capacities for listening are changed by technological culture. Its roots lie in the birth of the electronic communications revolution. Through this revolution we have learned to listen farther than any previous human generation. The telephone, the radio, and even the radio telescope have extended the range of our hearing as never before. It has also made technologically produced sound pervasive, as the Beatles and Beethoven alike blare forth from the living-room stereo.

But above all, the electronic communications revolution has made us aware that once silent realms are in fact realms of sound and noise. The ocean now resounds with whale songs and shrimp percussion made possible by the extension of listening through electronic amplification.

The distant stars, which perhaps are not so thoroughly in a "harmony of the spheres" of the Pythagoreans, nevertheless sputter in the static of radio-astronomy. In our urban environments noise pollution threatens the peace of mind which we now wishfully dream of in terms of quieter eras.

It is not merely that the world has suddenly become noisier, or that we can hear farther, or even that sound is somehow demandingly pervasive in a technological culture. It is rather that by living with electronic instruments our experience of listening itself is being transformed, and included in this transformation are the ideas we have about the world and ourselves.

If we grant that the origins of science lie with the Greeks, aided by the sense of mastery implied in the human role of cocreator with the Hebrew God, there remains a distinct distance from both Greek science and Hebrew theology in the rise of technology. Contemporary science is *experienced* as embodied in and through instruments. Instruments are the "body" which extend and transform the perceptions of the users of the instruments. This phenomenon may be considered apart from the usual considerations of the logic of the sciences, of the inner language of science in mathematics, and it may be investigated in terms of the experience through technology of the world, others and myself.[2]

What is of special interest to the thoughtful listener is then the way instruments, particularly those of the electronic era, introduce ways of listening not previously available. If one playfully turns to a speculative consideration of the role of instrumentation as a means of embodied experience in relation to the rise of modern science, a hypothesis suggests itself. Whether by historical accident or a long-held and traditional preoccupation with vision, the new scientific *view* of the world began with equally new instrumental contexts made possible by the emerging technologies of lens grinding and a concern with optics. Galileo's moons, never before seen, are *experienced through* the embodying and extending instrument of the telescope. The universe comes into view, is *observed* in its ever-extending macrocosm, through the instrument. Nor does it make any essential difference in the phenomenon of the transformed experience whether the discovery follows and confirms a speculation or initiates and inaugurates a new view of things. In either case what was previously unseen occurs within experience itself. The same occurs under the gaze of the microscope. A miniworld never before seen even if its existence had been suspected unfolds with a wealth and richness of animals, plants, cells, and microbes not dreamed of in the theoretical imagination which preceded the perception. Thus with increasingly passionate excitement humankind became more and more entranced with this extension of its vision.

Subtly, however, the extension of vision not only transformed but *reduced* humankind's experience of its newly found domains. For the picture of the world which began to unfold through the new instrumentation was essentially a *silent* world. The macrocosmic explosions of the stars and the microcosmic noise of insects and even of cells had not yet reached the human ear. If today we know that this silence was not a part of the extended but reduced world of early modern science, it is in part due to the later development of another means of embodiment through electronic instruments. What was first seen was later given *voice*.

In the gap between optics and electronics in this speculation, the sense of the world moved from the once silent Galilean and Newtonian universe to the noisy and demanding universe of today. But almost by rebound the intrusion of sound perhaps reveals something about our previous way of thinking, a thinking which was a viewing, a *worldview*. We have discovered a latent, presupposed, and dominant visualism to our understanding of experience. If on the popular front it has taken those concerned with media, such as Marshall McLuhan, and Walter J. Ong, to point this out for contemporary consciousness, it is because this visualism has long been there for us to see had we but the reflective power to discern it.

This visualism may be taken as a symptomatology of the history of thought. The use and often metaphorical development of vision becomes a variable which can be traced through various periods and high points of intellectual history to show how thinking under the influence of this variable takes shape.

The visualism which has dominated our thinking about reality and experience, however, is not something intrinsically simple. As a tradition it contains at least two interwoven factors. The first is more ancient and may be thought of as an implicit *reduction to vision* whose roots stem from the classic period of Greek philosophical thought. Its source lies not so much in a purposeful reduction of experience to the visual as in the glory of vision which already lay at the center of the Greek experience of reality.

In contemporary philosophy it has been Martin Heidegger who has made us most aware of the deeper roots of the vision of the Greeks. Through his radical analysis of the question of Being, Greek thinking itself emerges as the process of allowing Being to 'show forth' as the 'shining' of *physis*, of the 'manifestation' of Being as a 'clearing' all of which recalls the vibrant *vision* of Being. Nor is Heidegger alone in this recognition of the intimacy between vision and the ultimately real for Greek thought. Theodor Thass-Thienemann notes, "The Greek thinking was conceived in the world of light, in the Apollonian visual world . . . . The Greek language expresses this identification of 'seeing' and 'knowing' by a verb which means in the present *eidomai*, 'appear', 'shine', and in the past *oida*, 'I know', properly, 'I saw'. Thus the Greek 'knows' what he has 'seen'."[3] Even the Greek verb meaning "to live" is synonymous with "to behold light."[4] Before philosophy and deep in the past of Greek experience the world is one of vision. In this sense visualism is as old as our own cultural heritage.

But with the development of philosophy, more with its *establishment* in the Academy and the Lyceum, the preference for vision expressed in the wider culture begins to become more explicit. Visualism arises with a gradual distinguishing of the senses. One of the earliest examples lies in the enigmatic claim of Heraclitus that "eyes are more accurate witnesses than ears."[5] Not being given a context for the fragment, it is of course quite difficult to discern what Heraclitus meant. He could have meant that to see something happen in the flesh is more accurate than to hear of it through gossip. But even if this is not what he had in mind, the relation of sight and accuracy already appears to be established. Experientially it is not at all obvious that eyes are more discriminating than ears.

Even the ordinary listener performs countless auditory tasks which call for great accuracy and discrimination. In physical terms the mosquito buzzing outside the window produces only one-quadrillionth of a watt of power; yet one hears him with annoyance, even if one can't see him. And the moment trained listening is considered, feats of discrimination become more impressive. The expert auto mechanic can often detect the difficulties in an engine by sound, although when it has been taken apart the play in the bearings may be difficult to see. And in the paradigm of disciplined listening, the musician demonstrates feats of hearing which call for minute accuracy. The listener to the subtlety of Indian music with its multiple microtones discovers an order of extremely fine auditory embroidery.

But whether or not Heraclitus stated a preference for vision which may already conceal a latent inattentiveness to listening, Aristotle, at the peak of academic philosophy, notes, "Above all we value sight . . . because *sight is the principle source of knowledge* and reveals many differences between one object and another."[6] Here is a clearer example of a preference for vision and emerging distinctions among the senses.

Several features of this text stand out. First, it is clear that Aristotle notes that the valuation of sight is already something common, taken for granted, a tradition already established. Second, there is again the association of sight with differences and distinctions which may be the clue to a latent inattentiveness to listening. But, third and most important, the main thrust of Aristotle's visualism lies in the relation between sight and *objects*. The preference for vision is tied to a metaphysics of objects. Vision already is on the way to being the "objective" sense.

Once attention to the latent visualist tradition of philosophy is made concerning the intimate relation between light imagery and knowledge, a flood of examples comes to mind. For visualism in this sense retains its force in English and in most related Indo-European languages. Only the briefest survey shows the presence of visual metaphors and meanings. When one solves a problem he has had the requisite *insight*. Reason is the *inner light*. There is a *mind's "eye*." We are *enlightened* when informed by an answer. Even the lightbulb going on in a cloud over the cartoon character's head continues the linkage of thought with vision.

Less obvious but equally pervasive are the terms which, while they have lost the immediacy of light imagery, retain it at the root meaning. *Intuition* comes from the Latin *in-tueri*, "to look at something." Even *perceive* is often implicitly restricted to a visual meaning. Vision becomes the root metaphor for thought, the paradigm which dominates our understanding of thinking in a reduction *to* vision.

Philosophy and its natural children, the sciences, have often blindly accepted this visualism and taken it for granted. It is not that this tradition has been unproductive: the praise of sight has indeed had a rich and varied history. The rationality of the West owes much to the *clarity* of its vision. But the simple preference for sight may also become, in its very richness, a source of the relative inattentiveness to the global fullness of experience and, in this case, to the equal richness of listening.

Even within the dominant traditions there have been warnings in the form of minority voices. Empedocles called for a democracy of the senses.

> Come now, with all your powers discern how each thing manifests itself, trusting no more to sight than to hearing, and no more to the echoing ear than to the tongue's taste; rejecting none of the body's parts that might be a means to knowledge, but attending to each particular manifestation.[7]

And from the very earliest stratum of Greek philosophical thought Xenophanes voiced the note that experience in its deepest form is global: "It is the whole that sees, the whole that thinks, the whole that hears".[8]

Were, then, the dominant visualism which has accompanied the history of thought a mere inattention to listening, the praise of sound which may begin in its own way in the twentieth century would be but a corrective addition to the richness of philosophical vision. And that itself would be a worthwhile task. But the latent reduction *to* vision became complicated within the history of thought by a second reduction, a reduction *of* vision.

The roots of the second reduction lie almost indiscernibly intertwined with those arising from the preference for vision; the reduction *of* vision is one which ultimately separates sense from significance, which arises out of doubt over perception itself. Its retrospective result, however, is to diminish the richness of every sense.

For the second reduction to occur there must be a division of experience itself. This division was anticipated by two of the Greeks, Plato and Democritus, who were opposed in substance but united formally in the origin of Western metaphysics. For both, the ultimately real was beyond sense, and thus for both, sense was diminished. Both "invented" metaphysics.

This invention was the invention of a perspective, a perspective which was ultimately *imaginative*, but which in its self-understanding was the creation of a "theoretical attitude," a stance in which a constructed or hypothesized entity *apart from all perceptual experience* begins to assume the value of the ultimately 'real'. With Democritus the occasion for the invention of metaphysics came with the idea of the *atom*. The atom is a thing reduced to an *object*. Rather than a thing which shows itself within experience in all its richness, the atom is an object which has 'primary' qualities to which are added as effects 'secondary' qualities which are 'caused by' the primary qualities. Thus, too, is *explanation* born. The task of metaphysics is to "explain" how the division it introduces into the thing is overcome by a theory of complex relations between the 'primary' and the 'secondary' qualities.

Democritus's atoms are no longer things, they are "objects" which, while they may *seem* to possess the richness of things, at base are "known" to be poorer than things. Democritus's atoms, according to Aristotle, possess only *shape*, *inclination* (direction of turning), and *arrangement*. But note what has happened to sense: "visually" the atoms are "really" colorless, and insofar as they are colorless in "reality" they are "beyond" sense *in principle*. This is a leap which propels Democritus onto a path prepared for but never taken by his predecessors. Anaxagoras's "seeds," which were the predecessors of atoms, were *in practice* invisible, because they were too small for our eyes to see. What was lacking was a means of bringing them into view. But even though our powers are limited, for Anaxagoras "appearances are a glimpse of the unseen."[9]

But with the Democritean atom which is *essentially* colorless, what sense "gives" is placed under an ultimate suspicion. For Democritus it is "by convention that color exists, by convention sweet, by convention bitter." Knowledge is divided into sense, and what is not yet named but which is essentially different from sense. "Of knowledge there are two types: the one genuine, the other obscure. Obscure knowledge includes everything that is given by sight, hearing, smell, taste, touch; whereas genuine knowledge is something quite distinct from this."[10] This momentous turning was not taken without some doubt. Democritus heard this doubt in a voice given to the senses, "Ah, wretched intellect, you get your evidence only as we give it to you, and yet you try to overthrow us. That overthrow will be your downfall."[11] Nor is it ever clear that the "overthrow" succeeded completely. Even the atom retained one, though diminished, *visual* attribute in its *shape*. The preliminary result of the "invention" of metaphysics was the diminution *of* vision in its essential possibilities.

Plato in his own way made the same "invention." But Plato's version of the "invention" of metaphysics was, if anything, more complete than Democritus's. If Democritus's atoms retained one visible predicate, Plato's ultimate "reality," the Idea of the Good, in itself contained none but was presumably known only to the mind or intelligence. There does remain an *analogy* with the sensible, and that analogy is again visual. The Idea of the Good is "like" the sun in the visible realm. "It was the Sun, then, that I meant when I spoke of that offspring which the Good has created in the visible world, to stand there in the same relation to vision and visible things as that which the Good itself bears in the intelligible world to intelligence and intelligible objects."[12] But Plato steadfastly maintained that this was merely an analogy: "light and vision were thought to be as like the Sun, but not identical with it . . . to identify either with the Good is wrong,"[13] because the distinction between the visible or sensible and the intelligible which founds the doctrine of forms of Ideas has already separated sense from reason. The sensible realm in its "likeness" or analogy to the purely intelligible realm of the Ideas becomes a "representation" which indicates what cannot be sensed. In the notion of imitation, mimesis, and representation lies the direction which is counter to that of the polymorphic embodiments of experience, and which lays the antique basis for the more modern forms of the dualism of experience which pervade the contemporary era. The ancient sources of the double reduction of experience in visualism did not become clear or mature until the opening of the modern era. Modern visualism as a compounded reduction of experience is clearly notable in the work of Descartes where both the Democritean and Platonic anticipations meet to form the basis of modern visualism. Descartes unites and preserves the ambiguities of the diminution of the senses in his praise of the *geometrical method.* For Descartes the light and visual imagery has become metaphorical in a rather perfunctory sense: "Having now ascertained certain principles of material things, which were sought, not by the prejudices of the senses, but by the *light* of reason, and which thus possess so great evidence that we cannot doubt their truth, it remains for us to consider whether from these *alone* we can deduce the explication of all the phenomena of nature."[14] Thus in the rise of modern metaphysics there is retained the echo of a distrust of the senses and a corresponding faith in reason as an invisible, imperceptible realm of truth.

With Descartes the progression of the diminution of sense continues, and the object is now reduced to its geometric attributes: he further reduces the Democritean atom. "The nature of the body consists not in weight, hardness, colour and the like, but in extension alone it is in its being a substance extended in length, breadth, and depth."[15] Here the Democritean anticipations of a doctrine of 'primary' and 'secondary' qualities take the form of being defined in geometric terms. Extension is 'primary' and all other qualities are 'secondary' or derived.

But Descartes repeats the Democritean ambiguity. While claiming that "*by our senses* we know nothing of external objects beyond their figure, magnitude, and motion,"[16] his ultimate aim is a total denial of sense.

> "But, since I assign determinate figures, magnitudes, and motions to the insensible particles of bodies, *as if I had seen them*, whereas I admit that they do not fall under the senses, some one will perhaps demand how I have come by my knowledge of them. To this I reply, that I first considered . . . all the clear and distinct notions of material things that are to be found in our *understanding* . . . which rules are the principles of geometry and mechanics, I judged that all the knowledge man can have of nature must of necessity be drawn from this source."[17]

In spite of this extrapolated claim the now geometrically reduced object even at its insensible level retains certain "abstract" visual properties. However, the "real" object is now thought to be a bare and reduced object distinctly different from the rich thing found in experience.

What Descartes accomplishes, here using what happens to vision as a symptom for what happens to experience overall, is a division of experience into two realms so that one region of experience is made to rule over all others. The reduced abstract object (extended object) becomes "objective" and its appearance within perceptual experience with the significant exception of those ghostly remaining visual qualities becomes "subjective." Simultaneously reason, understanding, the geometrical deductive process, become disembodied as "pure" acts of mind.

Descartes's counterpart, John Locke, disagreed that the source of clear and distinct ideas was the understanding—it was rather experience —but in formulating the grounds of empiricism Locke preserved the ancient distrust of perception in a new way. Seeming to take seriously and to take account of sense experience, Locke ended by reducing it to a sense automism which again separated knowledge from things.

Locke, as did Descartes, perfunctorily maintained the metaphor between seeing and understanding. "The understanding, like the eye, whilst it makes us see and perceive all other things, takes no notice of itself; and it requires art and pains to set it at a distance, and to make it its own object."[18] But in Locke's case, if the metaphor was to be extended, it was not the eye but an outside influence which provided its own objects. Thus the classical empiricist thesis:

> Let us suppose the mind to be, as we say, white paper, void of all characters, without any ideas; how comes it to be furnished? Whence comes it to be that vast store, which the busy and boundless fancy of man has painted on it with an almost endless variety? Whence has it all the materials of reason and knowledge? To this I answer in one word, from EXPERIENCE.[19]

The door is opened in this thesis to things and the richness of experience, but Locke so quickly borrowed from Descartes the notion of clear and simple ideas that mundane experience was immediately bypassed for what became empiricist atomism. Locke believed, in an echo of the analytic and geometric prejudice, that what was primitive in experience had to be the simple, and thus the simple and already analyzed idea was in effect the object which was immediately before the mind in experience, "that term which, I think, serves best to stand for whatsoever is the object of the understanding when man thinks."[20] But such simples are better called *concepts* than perceptions, whereas perception for empiricism becomes the result of an unfelt and unexperienced pointillism of abstract qualities.

Locke paused only briefly before the things. "Though the qualities that affect our senses are, in the things themselves, so united and blended that there is no separation, no distance between them,"[21] he did not hesitate to immediately conclude that "yet it is plain the ideas they produce in the mind enter by the senses *simple* and *unmixed*."[22] These ideas which are simple and unmixed are the 'atoms' of sensory qualities, 'abstract' qualities apart from any thing. "Thus we come by those *ideas* we have of yellow, white, heat, cold, soft, hard, bitter, sweet, and all those which we call sensible qualities."[13] That no one has ever perceived a disembodied white did not seem to trouble Locke, and the empiricist tradition to this day debates the way we build up objects, and things from these simple ideas become 'sense data'.

Nor is this the end of the Lockean version of the reduction of the thing. Locke specifically enunciated the previously implicit doctrine of primary and secondary qualities, that is, of the various atoms of qualities some are privileged and others are mere effects of the privileged qualities.

Primary qualities were thought by Locke to be qualities *of* the material object (the reduced object). "Qualities thus considered in bodies are, First, such as are utterly inseparable from the body in what estate soever it be."[24] And these qualities remain cartesian and visual, although they are more complex than those allowed by Descartes (and allowing one quality which Locke thought belonged to *tactile* perception as well): "These I call *original* or *primary qualities* of body, which I think we may observe to produce simple ideas in us, viz. solidity, extension, figure, motion or rest, and number."[25] Secondary qualities are those "which in truth are not in the object themselves, but powers to produce various sensations in us by their primary qualities."[26] Thus Locke repeated in essential outline the metaphysical division of the thing which results in its reduction.

This division was already enough to establish the need for empiricism to face the problem of how the thing is built up from its simple atoms, but a second dimension to the division was also affirmed by Locke, the atomism of the senses. It is quite clear that in his interpretation of the already extant tradition of five senses, the senses had now become more "clear and distinct" so that some qualities enter experience from one sense only, and others enter from the other senses. Thus the thing remains, in itself, an object of primarily visual - spatial attributes to which in the mystery of experience are "added" the various simple and "subjective" ideas of other qualities. Both the thing and experience remain under the limitation of the double reduction.

This progressive march of reductionism in philosophy is more than a mere visualism which stands as its symptom. It is a tendency which lies more deeply in a certain self-understanding of philosophy. On a surface level, and again symptomatically, a visualism can be called into question by pointing up consequences which lead to the inattention to important dimensions of experience in other areas, here, in particular in an inattention to listening. Not only are sounds, in the metaphysical tradition, secondary, but the inattention to the sounding of things has led to the gradual loss of understanding whole ranges of phenomena which are there to be noted.

What is being called visualism here as a symptom is the whole reductionist tendency which in seeking to purify experiences belies its richness at the source. A turn to the *auditory dimension* is thus potentially more than a simple changing of variables. It begins as a deliberate decentering of a dominant tradition in order to discover what may be missing as a result of the traditional double reduction of vision as the main variable and metaphor. This deliberate change of emphasis from the visual to the auditory dimension at first symbolizes a hope to find material for a recovery of the richness of primary experience which is now forgotten or covered over in the too tightly interpreted visualist traditions.

It might even be preliminarily suspected that precisely some of the range of phenomena at present most difficult for a visualist tradition might yield more readily to an attention which is more concerned with listening. For example, symbolically, it is the *invisible* which poses a series of almost insurmountable problems for much contemporary philosophy. "Other minds" or persons who fail to disclose themselves in their "inner" invisibility; the "Gods" who remain hidden; my own "self" which constantly eludes a simple visual appearance; the whole realm of spoken and heard language must remain unsolvable so long as our seeing is not also a listening. *It is to the invisible that listening may attend.*

If these are some of the hopes of a philosophy of listening and voice, there remains within philosophy a strong resistance to such a task. For philosophy has not only indicated a preference for the visual and then reduced its vision from the glowing, shining presence of *physis* to its present status as the seeing of surfaces as combinations of atomized qualities, but it has harbored from its classic times a suspicion of the *voice*, particularly the sonorous voice. Although there may be a certain touch of irony in the *Republic* of Plato (who could be a more subtle rhetorician than Socrates?), the intimation of danger in poetry, dramatic recitation, and even in certain music remains. There is in philosophy a secret tendency toward a morality of sparseness which today is typified by a preference for desert landscapes. Socrates noted, "It strikes me, said I, that without noticing it, we have been purging our commonwealth of that luxurious excess we said it suffered from."[27]

In the wider Greek culture, however, the Apollonian love of light was balanced by the Marsyasian love of sound. The tragedies spoke in sonorous voices through the *persona*, or "masks," which later are held to mean also *per-sona* or "by sound." Nietzsche, who much later placed into a dialectic the Apollonian and the dark and furious Dionysian, affirmed that one must also accept a "god who dances" as well as the stability of Apollonian form. Yet in spite of the apparent domination of a new reduced Apollonian visualism, there is also another root of our Western culture which takes as primary a version of a "god who dances" with the movement and rhythm of sound.

That tradition is not that of philosophy but that of the Hebrew theology of the imagery of word and sound. The primary presence of the God of the West has been as the God of Word, YHWH. "And God *said*, let there be ———. . . . " The creative power of the Hebrew God is *word* which is spoken forth as power: *from word comes the world.* And although God may hide himself from the eyes, he reveals himself in word which is also event in spite of the invisibility of his being. Human life, too, as the word-breath which unites the human with others and the gods is a life in sound. But if the world is devocalized, then what becomes of listening? Such has been a theological question which has also pervaded our culture.

A theology is not a philosophy, and what is needed is not a revival of theology, not even a secular theology. For so long as the gods remain silent—and if they are dead they have fallen into the ultimate silence—no amount of noise will revive them. But if they speak they will be heard only by ears attuned to full listening. For what is needed is a *philosophy* of listening. But is this a possibility? If philosophy has its very roots intertwined with a secret vision of Being which has resulted in the present state of visualism, can it listen with equal profundity? What is called for is an ontology of the auditory. And if any first expression is a "singing of the world," as Merleau-Ponty puts it, then what begins here is a singing which begins in a turn to the auditory dimension.

But while such a symptomatology has its tactical uses, a deliberate decentering of visualism in order to point up the overlooked and the unheard, its ultimate aim is not to replace vision as such with listening as such. Its more profound aim is to move from the present with all its taken-for-granted beliefs about vision and experience and step by step, to move towards a radically different understanding of experience, one which has its roots in a *phenomenology* of auditory experience.

LISTENING AND VOICE

1. Philip Wheelwright, *The Presocratics* (New York: Odyssey Press, 1966), p. 60.
2. See Patrick Heelan, "Horizon, Objectivity and Reality in the Physical Sciences," *International Philosophical Quarterly* 7 (1967): 375–412. Also, Don Ihde, "The Experience of Technology," *Cultural Hermeneutics* 2 (1974): 267–79.
3. Theodor Thass-Thienemann, *Symbolic Behavior* (New York: Washington Square Press, 1968), p. 147.
4. F. David Martin, *Art and the Religious Experience* (Lewisberg, Pa.: Bucknell University Press, 1972), p. 236.
5. Wheelwright, *The Presocratics*, p. 70.
6. Aristotle, *Metaphysics*, trans. John Warrington (London: J. M. Dent and Sons), p. 51 (italics mine).
7. Wheelwright, *The Presocratics*, p. 70.
8. Ibid., p. 32.
9. Ibid., p. 160.
10. Ibid., p. 182.
11. Ibid., p. 182.
12. Plato, *The Republic*, ed. Francis Cornford (London: Oxford University Press, 1945), p. 219.
13. Ibid., p. 220.
14. René Descartes, *A Discourse on Method*, trans. John Veitch (London: Everyman's Library, 1969), p. 212 (italics mine).
15. Ibid., p. 200.
16. Ibid., p. 220 (italics mine).
17. Ibid., p. 225 (italics mine).
18. John Locke, *An Essay Concerning Human Understanding*, ed. A. S. Pringle-Pattison (Oxford: Clarendon Press, 1969), p. 9.
19. Ibid., p. 42.
20. Ibid., p. 15.
21. Ibid., p. 53.
22. Ibid., p. 53.
23. Ibid., p. 43.
24. Ibid., p. 66.
25. Ibid., p. 67.
26. Ibid., p. 67.
27. Plato, *The Republic*, p. 87.

CHAPTER TWO

UNDER THE SIGNS OF HUSSERL AND HEIDEGGER

The examination of sound begins with a phenomenology. It is this style of thinking which concentrates an intense examination upon experience in its multifaceted, complex, and essential forms. Nothing is easier than a "phenomenology," because each person has his experience and may reflect upon it. Nothing is more "familiar" than our own experience, nor anything closer to ourselves. Potentially anyone can do a "phenomenology."

But nothing is harder than a phenomenology, precisely because the very familiarity of our experience makes it hide itself from us. Like glasses for our eyes, our experience remains silently and unseeingly presupposed, unthematized. It contains within itself the uninterrogated and overlooked beliefs and actions which we daily live through but do not critically examine.

There is also a purposeful naïveté to phenomenology in regard to experience as it "returns" to that experience. But that naïveté is not a first or easy one. It is a second naïveté which arises out of a critical and controlled discipline of investigation. The first task of phenomenology is to replace the easy naïveté of ordinary reflection with the difficult second naïveté of phenomenology proper.

Behind the stance of phenomenology proper with its own rigorous naïveté there stands a history guided primarily by the philosophies of Edmund Husserl and Martin Heidegger. For my purposes I shall take these two founders of phenomenological investigation to belong to the same style of thought, although they both started from different questions. Husserl will be the guide for what may be called *first phenomenology*, while Heidegger will be the guide for a *second phenomenology*.

First phenomenology, initiated by Husserl, is precisely the working out of both a method and a field of study. As a method, self-consciously developed, the Husserlian phenomenology is one which is dominated by a highly technical language and set of intellectual machinery. *Epoché*, *the phenomenological reductions*, *bracketing*, and the various terms which go with Husserl are to be here viewed as a means of gradually approximating a certain stratum of experience. It is a beginning which, through both the deconstruction of taken-for-granted beliefs and the reconstruction of a new language and perspective, becomes a prototype for a *science of experience*.

Second phenomenology begins where first phenomenology leaves off. It takes for granted the attainments of phenomenological method in its most radical sense and directs its questions to both an extension and a deepening of the formal ontologies of Husserl toward a fundamental ontology of Being. Its aim is that of a hermeneutic and existential philosophy.

But if the beginning is one which opens as a "science" of experience and later as second phenomenology leads into the question of "existential" language, there remains in both phenomenologies a sense of learning where one feels even as he enters a new language that he has known it all along. Breaking with the easy familiarity of experience, deliberately putting it at a distance, leads to a return of enriched significance again "familiar" but also subtly changed. Phenomenology allows us to belong to our experience again but hopefully in a more profound way.

One secret of the singleness of the way between first and second phenomenology lies in the distance which emerges between the *center* of experience and things and the *horizon*. For Husserl the center of attention and of all experience is *intentionality*, that essence of experience to be directed towards, to be "aimed" at. And in first phenomenology the concern is to take note of, to describe, and analyze the ways that directedness takes place in both language and perceptual and imaginative experience. The things which are intended and the acts by which their meanings are constituted occupy first phenomenology centrally.

Nor are the things of the world ignored in second phenomenology, but once the center is discovered, the way is also opened "outwards" towards limits and horizons. It is increasingly this question which animates second phenomenology and is the source of its at first seemingly odd language. But the unity which lies between first and second phenomenology can be concretely discovered only along the way. Nor does one begin with the end.

There is a preliminary and simple way in which the relation between the two guiding figures of phenomenology can be understood, however. "To the things themselves"[1] was the worthy motto of the Husserl who meditated upon *epoché*, that turning of thought which creates phenomenology. The problem is one of beginnings; but that has been a perpetual problem for philosophers, because a beginning is always made in the midst of that which has already begun. To begin anew therefore calls for a new way of getting to the things and a new way of expressing that turn. Thus the beginning is one which is a certain struggle with language.

The strategy for beginning, in Husserl's case, was one which called for the elaboration of a step-by-step procedure through which one viewed things differently. His model was one of analogy to various sciences, often analytic in style; thus he built a methodology of steps: *epoché*, the psychological reduction, the phenomenological reduction, the eidetic reduction and the transcendental reduction. At the end of this labyrinth of technique what was called for was a phenomenological attitude, a perspective from which things are to be viewed.[2] In this, first phenomenology operated like a science and is in the first instance a *statics* of experience.

Historically it was once, after all, strange and unnatural to inhabit the imaginary standpoint in the Copernican view of the earth which called for an observer to imaginatively place himself "outside" the solar system and see an earth rotate around the sun in contrast to the earthbound "first" or naïve view which looked outward only to the sun circling the earth. But once the new viewpoint was made intellectually inhabitable, ever new discoveries became possible.

There is also a quality of new frameworks which, once having been learned, renders the once-necessary "machinery" not only easily operatable but makes it seem almost unnecessary. Once this language is truly learned, it need not be lived literally, since the way of seeing is the attainment. Thus those who followed Husserl, for the most part, abandoned the mechanism of the scaffolding and developed a more existential language with the intention of dealing with the edifice itself. Such was the case with Heidegger who implicitly follows Husserl's steps without explicitly noting each step of the method. The active attempt to grasp "the things themselves" becomes a "letting be" of the phenomena to "show themselves from themselves."[3]

Yet despite the connotations of ease, this new version of *epoché* turns out to be equally difficult. To let the *things* speak, to show themselves, calls for an "act" of special restraint on the part of the seeker which, in the case of Heidegger, is the gradual unlayering of the deeply entrenched traditions of thought which continue to enmesh the things themselves in the way they may be viewed or heard. Thus the "destruction of the history of ontology" is very much a part of the new *epoché* of second phenomenology. It understands that experience cannot be questioned alone or in isolation but must be understood ultimately in relation to its historical and cultural imbeddedness.

The distance and relation between first and second phenomenology are reflected in the preliminary results of each. First phenomenology often yields an early appreciation of the *richness* and complexity of experience. But second phenomenology in pursuing that richness discerns in the sedimentation of our traditions of thought an essential embedment in *history* and *time* of experience itself. For while the first word of phenomenology is addressed to the nearness of experience as a philosophy of presence, second phenomenology is a rebound which opens the way to a reevaluation and reexamination of the very language in which our experience is encased and by which it is expressed. The phenomenology of essence, structure, and presence in Husserl leads to the phenomenology of existence, history, and the hermeneutical in Heidegger.

But in actuality the opening to the world which is phenomenology is simultaneously both. The rediscovery of the richness of experience and its structures is a discovery of the essential embedment of experience in historicality and therefore in the polymorphous flexibility of human being. It is not accidental that historically Heidegger's *Being and Time* and Husserl's *Crisis* are the most similar works of the two authors pointing to a convergence of the two phenomenologies.

My purpose here, however, is not to digress into the history of phenomenology but to *do* a phenomenology in the light of its past. By distinguishing two phenomenologies which ultimately belong together, a movement is initiated which begins in *approximations*. I shall begin the inquiry in a Husserlian-styled first phenomenology and by approximations move toward a more existential philosophy of listening and voice.

This beginning in approximations itself reflects the historical movement of phenomenology in that the first approximations are "abstract" and not fully existential. Yet it is precisely this tendency to accept certain "abstractions" about experience which is closest to the sedimented traditions of thought which I wish to question. The approximations are therefore deliberate. They move away *from* the implicit acceptance of some ordinary and commonsensical understandings of experience towards a more vigorous understanding. This is particularly the case with what may be discerned as a kind of functional

gradual discovery of the possibilities of a genuinely *descriptive ontology*. First phenomenology calls for a thorough reinterpretation of common sense and science from its own insights. The unfortunate belief that phenomenology is thus "anti-scientific" or even counter to ordinary experience is a confusion held equally by some phenomenologists and those who would maintain that science is impossible without its "Cartesian metaphysics." But this is a connection which is not considered as essential by other phenomenologists perhaps more thoroughly cognizant of the sciences.

Until this confusion is sorted out—and it can be sorted out only *after* the phenomenological perspective is clearly gained—there must remain within all proximate parts of first phenomenology a certain polemic against Cartesianism and metaphysics at least in a heuristic way.

It is after gaining a certain grasp upon experience taken phenomenologically that the implications of the polemic begin to show the need for a second step, a second phenomenology. Thus through the approximations and a first movement which is styled after the descriptive phenomenology of Husserl, I shall begin a second movement which allows descriptive phenomenology to make its transition to the *existential*. It is *through* descriptive phenomenology that the existential dimension is first grasped in its significance. The existential is not a return to the "natural attitude," although existential significations function as the ultimate "natural attitude" of phenomenology. The problem will be to show in the process how the more radical language of a Heideggerian-styled second phenomenology is "natural" once its proper location is detected, for second phenomenology remains *descriptive*. This, however, must also be shown in the process itself.

The subject matter of this double inquiry is the whole range of auditory phenomena. This is to be a phenomenology of sound and listening. Beginning with an inquiry into the structures and shapes of sound, into existential possibilities of auditory experience, the investigation will range across a wide variety of human experiences in which sound and listening play crucial roles.

Clearly with humankind there is that focal speaking and listening activity of our babbling being, language in its auditory form. Closely related to spoken and heard language is the range of musical phenomena. There is also the noise and voice of the environment, of the surrounding lifeworld. There is the enigma, particularly for a first phenomenology of presence, of the horizon of silence. In more existential terms, the voices of language, of instruments, of the earth implicate things, persons, and the gods. For listening is listening to ———. And pervading the whole of the auditory dimension is the question of the inner voice as well. Each of these items must be queried in such a phenomenology.

1. See, for example, Edmund Husserl, *Cartesian Meditations*, trans. Dorion Cairns (The Hague: Martinus Nijhoff, 1960), p. 12; or his *Ideas: General Introduction to Pure Phenomenology*, trans. W. R. Boyce Gibson (New York: Collier Books, 1962), p. 49.
2. Husserl, *Ideas*, p. 39.
3. Martin Heidegger, *Being and Time*, trans. John Macquarrie and Edward Robinson (New York: Harper and Row, 1962), p. 51.

CHAPTER THREE

FIRST PHENOMENOLOGY

First phenomenology begins under the sign of Husserl. It is a beginning which strives to move us from where we are in terms of common assumptions and implicit beliefs to a different plane of understanding in the phenomenological attitude. To do this, a philosophy in the *style* of Husserl employs a double level of meanings. The first level of a Husserlian styled philosophy may be termed "literal"; and at this level of meaning phenomenology may be understood as a *philosophy of experience*, but a philosophy constructed along the lines of most previous philosophies.

Thus the phenomenological "metaphysics" is one which is based on what has been often called a *radical empiricism*. At this first level, phenomenology, sometimes characterized by Husserl as the creation of a genuine and *pure* descriptive psychology[1] is a science of the mind in contrast to the sciences of physical extension in the Cartesian paradigm. The aim is to isolate, describe, and discern the structures of immediacy or of fulfillable *experiential presence*. It is this aim which in retrospect can be seen to determine the theory of evidence which emerges in and from the phenomenological investigation itself. Primary evidence, "primordial dator evidence," as Husserl termed it, is anything that can actually be noted within experiential presence in the way in which it gives itself out. It is around this primary evidence that all other evidences must be scaled, judged, and arranged. In this, first phenomenology is a *philosophy of presence*.

Again, in the first and literal reading of Husserlian-styled phenomenology, this primary evidence is considered as that which is "given," although the "givenness" may not be what is only superficially present at a first glance. The task of isolating the appearance of experience as a phenomenon is actually an imposing one.

Secondly, in its first form, phenomenology is basically a statics of experiential presence. It seeks to uncover, once the field of primary evidence is isolated, the structures, invariants, and essential possibilities of that field. And while it turns out that not all essences will be clear and distinct (some will turn out to be "inexact," or, in Wittgensteinian terms, will have "blurred edges"), the aim of the first form of phenomenology is to make as precise as possible the shape of the experience being investigated.

In both these moves, first phenomenology as a *radical* type of "empiricism" remains roughly within the framework of traditional metaphysics, at least so long as the reading is literal. Primary evidence, fulfillable experiential immediacy, forms the foundation or ground-stratum as a given from which all other evidences derive or to which they relate back. Thus one might say that the distinctiveness of this "empiricism" is that it makes *experiential immediacy* the ground-stratum or primary substance of its metaphysics (rather than mind, matter, or something else).

Of course phenomenological "empiricism" also differs from classical empiricism from the outset, for it turns out that its field of experience is ultimately *total.* Its radicalism lies in the way this totality is taken. For example, in perceptual phenomena as one (privileged) *region* of experience, there is no distinction between primary and secondary qualities. All qualities are from the first "horizontalized" and must be taken as they give themselves out. And what is discovered is that what is given *in* perceptual experience is not at all a set of discrete qualities, but *things*, precisely those rich things which "blended" and "unified" the Lockean qualities which Locke overlooked as primary.

But concepts also give themselves out within experience. "Red" or "white", however, must be investigated *in the way* in which they are given. One can experience a concept, for example in using it in a proposition, without seeing red at all; but that type of experiential presence is quite distinct from the "bodily presence" of a "red" thing as a perceived thing. Moreover, the task of discerning the relations between the experience *of* a concept and its relation to the embodied thing experienced perceptually is also a problem for phenomenology.

It is here that the discrimination of "distances" arises. For once the types of evidence are sorted out, the next task becomes one of arranging and relating all matters by their relations to the primary fields of "givenness". This is a matter of philosophical measuring or situating of regions of phenomena.[2] With this consideration a third relation to empiricism may be seen, for despite a radically different way of interpreting perception, first phenomenology with both Husserl and Merleau-Ponty is at least implicitly perceptualist. The focus of primary evidence is *perception*. Thus one set of distinctions internal to phenomenology arises over the distance and difference between fulfillable perceptual immediacy and any "higher" level and more distant "constructions" in other modes of thought. There is a weighted beginning in the concreteness of perceptual experience.

All of these complex considerations meet in a rather oversimple illustration. Consider a problem having to do with a language puzzle about color. In the mid-twentieth century it is quite often the case that if a professor asks his class, *is* black a color? he will be likely to receive a number of negative answers. As reasons for their answers the students will recite what they have learned concerning color from the sciences, perhaps claiming that "real" color is defined in terms of wavelengths of light. But if the professor's question is, What color is that? while pointing to the blackboard, the overwhelming answer will be, simply, black. What, then, is "black" "really"? Here the answers soon may become enmeshed in metaphysical commitments and arguments.

The phenomenologist, however, approaches this problem somewhat differently once his reductions have been put into play. His task is to locate the *difference* and the type of *distance* between these two meanings of color, or, to use his terminology, he seeks to know how these meanings are *constituted*. That there is a rather large difference of context is quite clear, but contexts, the phenomenologist claims, must always have some weighted focus to make the discernment of distance possible. For first phenomenology that weightedness is *presence* or experiential immediacy. Thus only the second case yields the *experiential* context for black *as* a color. Nor is it a mere color. A more profound analysis would reveal that constellated with the meaning "black" taken perceptually there are "values," "symbolic significances," and "feelings." In the other "physical" context the experience of color *as* color is irrelevant: in its purest sense there is *no* color as color experience, but a reading of an instrument. In the phenomenologist's terms, "physical" color is quite distinct from "perceptual" color.

That is because these two meanings are constituted differently. "Physical" color is "experienced" *through* a machine, it is "read" hermeneutically. "Physical" color is thus constituted by a certain instrumental context or use while "ordinary" or "perceptual" color is constituted by the immediate perceptual context. But what is important to note in this first illustration is that phenomenology seeks to note and clarify the distance and difference of these two sets of related phenomena in terms of the key value given to perception as the weight which allows all other values to be "measured" from it. In this, phenomenology does not appear to be vastly distinct from some contemporary analytic philosophies, except that phenomenology takes as its primary evidence the region of fulfillable experiential immediacy as a starting point.

However, the student first entering phenomenological studies must be wary, precisely because the first literal reading of Husserlian philosophy is in fact only a preliminary approximation to phenomenology in its full sense. This is already discernible to the careful reader of Husserl. Not only does he use familiar, or "empiricistic" language deliberately to *lend* a certain initial clarity to the enterprise (because clarity is closely related to familiarity) but he makes one aware that he is employing this familiarity heuristically by his use of quotation marks and metaphors. Ultimately a second, nonliteral or hermeneutic reading of phenomenology is necessary if the outline of genuine phenomenology is to be reached. For although phenomenology may begin by an apparent "pure psychology," Husserl maintained over and over again that this "pure psychology" was ultimately the way to *transcendental philosophy*. And once this step is taken one can see retrospectively that in its depth phenomenology is not a metaphysics at all, nor even an "empiricism," for its destiny is that of an existential-hermeneutic philosophy which arises out of a descriptive ontology.

This movement from a "literal" to a hermeneutic reading of first phenomenology comes while it is underway. And the entry to the venture arises with the construction of the Husserlian "machinery" which in this context has been simplified and stylized. The doorway to phenomenology is *epoché* and the doctrine of the *phenomenological reductions*.[3]

To shorten and simplify matters one may regard the various reductions as hermeneutic or operational "rules" under which phenomenology operates. *Epoché* is the initial and general term for the phenomenological reduction overall. In Husserlian terminology *epoché* includes "bracketing," the "psychological reduction," or "phenomenological reduction."

The term *epoché* in its broadest sense means "to suspend" or "to put out of play." But *what* is suspended is to be a certain set of taken-for-granted beliefs. It is a suspension of "presuppositions" rather than a reduction of (primary) experience. But, understood here as a hermeneutic rule, *epoché* is an exclusionary and selective process. It is a rule which *excludes*, "brackets," "puts out of play," all factors which may not be noted as "bodily present" or actually fulfillable (intuitable) within ongoing experience.

This exclusionary rule is meant to place out of bounds certain ordinary and certain scientifically theoretical and even certain logical philosophical considerations. Thus, for example, in the descriptions of experience which are to follow, *epoché* as a rule excludes any physiological, physical "explanations" from the description itself. Also, all those assumptions of ordinary explanation must be put out of play, because ordinary experience already contains the sediments of "metaphysics." Furthermore, argument which infers or which draws upon some mediate line of reasoning is suspended in the interests of isolating and allowing the phenomenon to "appear" in its fulfillable or intuitable presence.

This exclusionary aspect of the rule is matched by its positive selectivity which sets apart for investigation the chosen field of experiential immediacy. It is to this field that phenomenology first turns. *Epoché establishes the "phenomenological attitude" or the perspective from which experience is to be taken.*

In this simple way this is quite easy to call for, although in actual practice the use of *epoché* turns out to be exceedingly difficult. But in its very difficulty one learns about the aim of phenomenology. Extensive work and thoughtful effort are called for in establishing the purposeful and disciplined phenomenological "naïveté" which is the beginning of phenomenology.

The second hermeneutic rule applies to the field of fulfillable experience, the selected and isolated field of investigation. It must be strictly correlated with the first rule: *Describe the appearances or phenomena.* In this case description calls for a careful note taking of what goes on in the "flow of experience." Moreover, the descriptions undertaken presuppose the "purification" called for in the first rule: Describe, don't explain.

In Husserl's own works there are constellated around this second rule a series of other subrules. For example, in *order* of concern, once a field for investigation has been selected, one begins with the "objects," or things which are "out there." These were called the *noema*, or "object-correlates," of the experience process.

But as these terms will be better introduced later, the way of description may begin partially here. At first a field of investigation will appear to be confusing, precisely because there are too many features to be noted. But what is of interest in the investigation is the *eidetic* or structural components of the experience in question. Yet these "patterns" are not immediately apparent. In this, again, phenomenological "psychology" is like any new science. It must look again and again at the phenomenon before it reveals its secrets.

There are two ways in which structures or invariants may begin to appear. The first may be called cataloging, which is the crude taking account of what goes on by listing what shows itself in a given moment of the "flow of experience." It soon becomes apparent that this list will be immense. But it does serve to demonstrate the complexity of the phenomenological field. It also serves to broaden phenomenological attention which begins to take note of much that is merely implicit in ordinary experience.

Equally or even more effective in discerning structures is the "gestalt" occurrence. It may and often does occur that a single experience will show an essential structural feature. For example, if I suddenly snap my fingers to the side of someone who is sitting with eyes closed, the *essential* directionality of experienced snapping sounds is presented.

Both of these devices fall within the use of what Husserl termed the use of *variations*. Fantasy or imaginative but also *perceptual* variations are the main methods for detecting essences. In Husserl's case these variations were largely modeled upon the notion of imaginative variations in the logical and mathematical *essential* sciences. But for reasons which again will become clearer later, Merleau-Ponty preferred the use of actual *perceptual* variations, not only because he was more explicitly "perceptualist" than Husserl, but because often the wildest imaginations do not yield many of the possibilities of the perceptual world.

Beginning under the sign of Husserl the preference for the essentially possible over the factual must also be noted. Logic and mathematics are sciences in which the essential or possible takes precedence over the factual. Husserl's belief was that such sciences are in a sense "regional," that is, they do not exhaust the full range of possibilities. Husserl's hope was that phenomenology would create the ultimate essential science as the analysis of all of the various "regions" of possibilities. This was descriptive psychology become philosophy.

What is to be of special interest here is the notion of a particular type of essential possibility, one which relates to the dimension of auditory experience. Because of its limitation to a dimension of experience I shall term this region of possibility a region of *existential possibilities*. But because there is also a need for a preliminary and at first schematic outline of existential possibilities, it may be necessary to differentiate them from the more familiar "logical possibilities" of contemporary philosophy. Existential possibilities form a particular type of possibility in the investigation of an actual dimension of human experience. In particular the various uses and roles of imagination in the development of both logical and existential possibilities call for initial attention.

There is a sense in which it can be said, again particularly in relation to certain types of imagination, that philosophy has always used fantasy as its tool. On the contemporary scene, particularly since Husserl's day in the analytic philosophies, a sophistication of one type of imaginative variation has flowered, that of the logical possibility. But precisely because this type of possibility is now familiar, the potential for a serious confusion concerning the Husserlian wider use of imaginative variations arises. For Husserl and for all phenomenology, logical possibilities are but one dimension of possibility, in this they are "regional" in a certain sense.

Such a claim might at first seem outrageous; yet Husserl himself often enough made it by noting that logic must itself be one of the sciences which is to come under *epoché*.[4] But the purpose here is not to enter into an argument concerning that development and its implications. It is rather to open the way to a contrast of logical and existential possibilities such that a different type of variation which becomes greatly important in phenomenological "psychology" can emerge. But as it is understood within phenomenology, the existential is not a matter of mere "contingency." Yet that is the understanding which invariably arises from the logistic prejudices of the still positivistically inclined thinker familiar with "logical possibilities." "Contingent" possibilities must fall under the domain of psychology. Be that as it may, the existential possibility to be discussed in what follows has as its central demand that it be a phenomenon which can be fulfilled by an intentional aim, a phenomenon which is *experientially possible.*

It may be that the experientially possible is a "narrower" region than that of the logically possible, although until investigation it is also possible that the reverse is true. But in any case the first demand for the location of an existential possibility is that it be fulfillable in experience.

Not all (emptily) imaginable possibilities are in fact fulfillable. There is a whole range of presumably easily "imagined" possibilities which are deceptive in appearing to be "clear and distinct" when, upon closer examination, they turn out to be confused and incomplete. Thus in the isolation and description of an existential possibility it is necessary that it be checked *as* fulfillable. In this, looking for existential possibilities is like an empirical procedure in that each item is to be checked and verified in principle in actual experience.

In all the examples which follow, particularly in those which I have chosen to put forth stylistically as the autobiographical "I can———," the experiences have actually occurred. There is a sense in which phenomenology begins with the first person, *I.*[5] But such is not the last word. In *every* case the use of the stylistic "I can———" in this book has been checked against the experiences of others. There is also the possibility of an intersubjective cross-checking, correcting, and expanding of discovery of essential possibilities in phenomenology.

That an existential possibility be actually intuitable in experience is a necessary but not sufficient condition of its location. A second aspect of the description of such possibilities is that one must note carefully *how* and *in what way* the particular variation occurs. For example, is this an ordinary or an extraordinary phenomenon? Does it occur centrally or peripherally? Can it be pushed further? And so forth. This is the way that the quite different uses of variations as plays of fantasy and as other types of variations begin to be clarified. "Logical possibilities" call for a certain "abstractive" and "reductive" use of imagination. But existential possibilities ever more closely approach the concreteness of what is essentially human. Thus whereas I can emptily "imagine" or conceive of a "world" of sound as a "No-Space" world in Strawson's sense,[6] when I turn to all the variations of my fulfillable experience of listening I find this is essentially *false*. For such a "No-Space" experience to be actualized I should have to be disembodied—but then would there be any "hearing" at all? (Ultimately, were this the point at issue I should argue that it is essentially *impossible* to fulfill even the imagination of a "No-Space" world. Such a world is, in Husserl's term, an "empty intention"—but so is a square circle an "empty intention." What is being confused in such an "empty" imagination is a region of thought which might be called *supposing*, but supposings are not necessarily fulfillable. They, too, need to be investigated as a region of experience, but a too-quick leap from an empty supposing to some kind of existential possibility leads to the confusions of analytic empiricism.)

From what has been said to this point one may gather that the procedure of locating and determining existential possibilities is not argumentative in the usual philosophical sense. There is not to be found here an argument in the sense of a deduction or one in the form of hypothetical-deductive reasoning. There is rather a gathering of descriptive characteristics in relation to the region of experience being investigated. However, I should say by way of anticipation that such a gathering, particularly in its mosaic accumulation, plays within phenomenology a role which *functions like* an argument. The detection and descriptive analysis of some feature of experience may be thought of as an *intuitional demonstration.* I first perform the act which is called for and find or do not find the case to be such and such. In turn, I may call for another to perform the same operation in cross-checking the result.

There is, in this process, even the possibility of correction. The other may have noted something which I either did not detect fully or which I did not think so important and thus left out. However, like an argument, the condition for the possibility of cross-checking depends rather thoroughly upon both investigators holding to the same framework or perspective from which the demonstration may be sought. In an argument if both sides do not hold to, say, the law of the excluded middle, a vast confusion is likely to result. So, also in phenomenology, an intuitional demonstration depends, for accurate results, on a certain awareness of the "rules" of the procedure.

Fortunately, in a very general sense, the phenomenologist can rely on a certain latent "phenomenological" ability on the part of others just as the logician can depend, in a very general sense, on a latent ability on the part of others to learn what is necessary to conduct a correct argument. Thus I can rely preliminarily upon the other to have such and such experience and upon the other to be able to detect whether such and such may or may not be the case. But it remains important that all variations be checked and cross-checked and not taken in their first and most superficial sense.

Another example of possible different directions for the use and type of imaginative variation may be illustrated by the "habitual" differences between a philosopher's way of "seeing" things and that of an artist. For instance, a long and ancient tradition in philosophy involves the use of an abstractive imagination in the constitution of logical "essences" or universals such as *predicates*. Although there have been centuries of argument about the status of such entities, one illustration revolves around two different types of possibilities of perceptual experience.

If I place before myself a "white" duck, a "white" chair, and a "white-haired" old lady and assume the usual context of the philosopher's way of "seeing" the world, I will probably structure the situation by asking what is *common* to these three "objects," and probably I will quickly come up with "whiteness" or some such conclusion. Yet, in a critical examination it is not at all clear or "obvious" that there is this commonality as perceived, for were I an artist I might well note that the "white" of the duck is a soft, feathery white in its concreteness; the white of the chair is glossy, hard; and the string gray with white of the old lady's hair all strike me as "vastly different." Does the philosopher "overlook" the concreteness of the various whites? Or does the artist not attend to abstract universals? Yet the artist also has his own "essential" insight into the various whites as he makes them "shine forth" in his painting. Phenomenologically, the philosopher and the artist experience or focus their experiences in quite different ways in relation to the phenomena; yet also phenomenologically under the stipulations of *epoché* both are "equal," and there is a matter of too much rather than too little "truth." The problem is one of discerning different types of essential possibilities.

There is a sense in which Wittgenstein in particular was sensitive to such nuances of differences in a very "phenomenological" way. The notion of family resemblances, already noted as a counterpart to the Husserlian notion of some types of inexact essences, is an attempt to recognize the noncommon relatedness of many phenomena in the mesh of ordinary language which does not display simply some clear "logical" structure.

But the point here is not to trace the history of the polymorphy of conceptuality. It is to caution about confusing two directions of the use of fantasy variations in which constructed examples, often employed early in what follows, become increasingly suspect if the sense of the existential possibility is to be elicited.

To begin with, then, any fulfillable possibility of the dimension of experience being investigated may be considered an existential possibility of that dimension of experience. And in relation to a given dimension of experience there may be innumerable such possibilities. But these gradually reveal a shape. As existential possibilities are discovered, a map of the terrain of experience may begin to appear in rough outline. Then two further steps may be noted. First, there is the need to fill in and precisely define the outline; and, second, there is a need to try to discern the limits of the region. It is in relation to limits in particular that a special problem arises.

There is a sense in which any fulfillable intention attains an "apodicity" or certainty. In Husserl, "apodicity" or certainty is the *weaker* category. What is stronger is *adequacy*, but there is often a serious uncertainty about what can constitute adequacy. When has one truly reached the limits of an existential possibility? That doubt must remain here as well as in the realm of logical possibilities. All essential sciences seem to display a certain openness or infinity of direction.

However, what can be done with beginning variations is a depiction of a "mosaic" map or survey of the terrain into which the investigator enters. The sampling of some existential possibilities gradually builds up an outline. As gestalt psychologists know, a "mosaic" of dots is often sufficient to display the picture. I do not claim to have in any way exhausted or even to have reached totally adequate limits of a phenomenology of sound and listening in the descriptive section. But I would hold that the existential possibilities elicited are suggestive of a need for philosophy to examine human experience more deeply than it often has.

In this simplified way first phenomenology remains within what at first may be called a "pure" or "descriptive" psychology. *Epoché* which outlines the field and the descriptive demand with its tool of variations for the noting of essential possibilities form the first steps of a Husserlian-styled phenomenology.

The third step of the reductions, Husserl's "transcendental reduction," is what makes a phenomenological psychology philosophical. It allows the field of experience to become universal under the notion of *intentionality*. However, again for simplicity, the "transcendental reduction" may also be regarded as a hermeneutic rule, this time described as a correlation rule.

First phenomenology contends that once underway *all* experience, whether fulfilled or remaining "empty," is found to have a specific shape in that all experience is "referential," "directional," and "attentional." All experience is *experience of* ———. Anything can fill in the blank. The name for this shape of experience is intentionality.

But as a hermeneutic rule of correlation, intentionality may be seen to *function* in phenomenology by giving a shape to phenomenology itself, a "model" or paradigm for its understanding. There were different ways in which Husserl characterized this structure of experience. It was the ego-cogito-cogitatum, the self experiencing something. In later phenomenology this notion was purged of its Cartesian overtones and made into Being-in-the-World. It was also, with Husserl, the structure which, within itself, could be differentiated according to the sides of a relation. But throughout it is the *relationality* of intentionality which must be maintained if phenomenology is to remain phenomenology proper.

In the period of the *Ideas* the distinctions which later were modified in various ways set the pattern for intentionality. Within experience overall there is that which is experienc*ed*, that called the *object-correlate* or *noematic* correlate. And, in strict correlation with the *noema*, there is the act of experience or the experienc*ing* which was the "subject-correlate" or the *noetic* act.[7] Here, as a correlative rule, it is maintained by intentionality simply that for every object *of* experience there is an act or "consciousness" which apprehends that object, and for every act there is an "intended" correlate, although some may not be fulfilled (empty).

This correlation as the phenomenological "model" gives phenomenology its characteristic shape. Anything outside the correlation lies suspended under the previous terms of *epoché*. Thus any object-in-itself and equally any subject-in-itself remains "outside" phenomenology. It is here that the Husserlian avoidance of "realism" and "idealism"—both of which are ultimately inverse sides of the same "metaphysics"—arises. "Objectivism" and "subjectivism" are both part of a "Cartesian," dualistic myth to which Husserlian phenomenology sees itself opposed as the radical alternative.

To interpret this correlation rule into its simplest terms, I shall employ a modification of the later terminology of "Being-in-the-World" which may be illustrated as follows:

$$\underset{(a)}{\text{Human}} \xrightarrow{\quad (b) \quad} \underset{(c)}{\text{World.}}$$

In this diagrammatic scheme for the phenomenological correlation, (b) stands for the constant rule of relation between (a), the human experiencer, and (c), the experienced environment. The correlation (b) symbolizes intentionality as first directional (the direction of the arrow) *towards* the world (c) which may be taken either as an "object" or as the totality of a "surrounding" world. In its strictest sense although (c), as the noematic correlate may be described "first" and in a sense comes "first" in experience, it is never separate from its being experienced (a) in retroreference to the experiencer. The opposite is also true. Furthermore, the modality and type of experiential relation (b) is from the beginning variable and complex in that it includes all possible experience rather than just cognitive or judgmental experience.

In the first instance the human of the correlation is always "me." I am the first instance of the correlation. However, one important modification must be introduced from the outset to avoid the problem of a "transcendental illusion." Although the diagram simplifies, it also confuses, because it tempts one to view it "from the outside" or "from above." I who am "in" the equation am suddenly also "outside it" as a "transcendental ego."

Most post-Husserlian phenomenologists have rejected this interpretation, and if carefully understood the diagram has an interpretation which calls for no outside "transcendental ego." That interpretation is one which calls for a complementary aspect to the correlation of Human → World. If the "outward" facing arrow of the intentional relation symbolizes my primitive involvements with the surrounding world, I also find that I may *reflect* upon that involvement by way of a modification of experience. Yet this reflectivity is implicit in every experience as well. It may be symbolized by an "echo" arrow:

$$\overset{(a)}{\text{Human}} \underset{(b')}{\overset{(b)}{\rightleftarrows}} \overset{(c)}{\text{World.}}$$

By this if (a), the subject, is related primitively in (b), intense involvement, with (c), the surrounding world, (b') is the reflective "stepping back" or "distancing" which I may make *within* the larger context of involvements. Reflection (b') is a special mode of (b) as self-awareness of the primary experience. The implication—again quite properly "anti-Cartesian" in the phenomenological radical alternative—is that I do *not* "know myself" directly in Cartesian fashion. What I know of myself is "indirect" as a reflection *from* the world. This also applies to others: I know myself as reflected from others. What is primitive is the "immersion" in the world, as Merleau-Ponty puts it.

Reflection is, in a sense, an *experience of experience*, but even here it can be seen that as a reflective experience it retains the essential shape of intentionality as experience *of———and* implies that my own self-knowledge remains essentially *hidden*. To truly "know himself" in phenomenology one must "know the world." Reflective knowledge is, in spite of the present necessary linguistic conventions, quite distinct from a "Cartesian" introspective procedure.

With this highly schematized and minimal set of methodological notions it becomes possible to gradually take an investigative stance. In so doing it is possible to begin to take note of some features of perceptual experience and simultaneously to introduce secondary notions while under way. I have deliberately chosen to begin with visual experience and, further, to deliberately fashion some of the subsequent explications of intentionality upon a visualist paradigm. In the following description, the concern is with noting preliminary features of perceptual intentionality.

I sit at my desk composing this chapter. I pause to reflect upon the ongoing experiences just previously lived through, or even currently going on. I am perhaps first overwhelmed by the complexity and polymorphous character of that experience. Were I now to begin to catalogue, item by item, even what I might recall in the few moments just past, and were I to do this thoroughly, it is conceivable that a rather large list would result.

However, I decide to deliberately focus my attention upon one dimension of that ongoing experience, its visual dimension. I begin to take stock of what I see and how I see within the moment. I note that while composing I focus upon the words taking shape under the keys of the typewriter: I note the errors, the stylistic and grammatical oddities, curse the wrong letters. I focus my attention even more narrowly and note that it is fixed roughly to a certain area upon the paper. At this stage I may describe this as a certain area of relative "clarity" which "stands out" and which gradually "shades off" into an area of relatively less clarity or even further out to a certain "fuzziness." Noematically within my visual field there is a "center" of the clearly and distinctly perceived which shades off into a "periphery" or "fringe" of the indistinctly perceived.

Also, strictly co-present with this seeing in which I am involved there is what I recognize, however implicitly, as the "mineness" of the experience. My seeing, my attending, my focusing, or equally put, the phenomena which "stand out" refer back to "me."

I further note at this preliminary stage that there is a play of inversions which goes on in the "flow" of experience. When I am attending to the paper there is a "ratio" of what is clear to what is fuzzy which runs from center to periphery. And if I turn my eyes upward to note something typed farther up the page, what has just been seen as clear now becomes less so, but the ratio between the central and relatively clear focal area and the peripheral and relatively fuzzy peripheral area remains constant.

Here are, despite their familiarity in gestalt psychology and even in much common parlance concerning the visual field, the beginning outlines of a "structure" within visual experience, which is for phenomenology an intentional structure. The noematic *core* or area of *focus* of the visual "world" is preliminarily distinguishable from its noematic *fringe*. Correlatively, the act of attention *is* a focusing (noetic act) which as an experiential structure displays a central awareness which shades off into the barely aware or implicit consciousness which is at the "fringe" of more explicit or focused attending.

I return to my visual experience. I now note that ordinarily I am concerned with, focus my attention upon, things or "objects," the words on the page. But I now note that these are *always* situated within what begins to appear to me as a widening field which ordinarily is a *background* from which the "object" or thing stands out. I now find by a purposeful act of attention that I may turn to the field *as* field, and in the case of vision I soon also discern that the field has a kind of boundary or limit, a *horizon*. This horizon always tends to "escape" me as I try to get at it; it "withdraws" always on the extreme fringe of the visual field. It retains a certain *essentially* enigmatic character. But within the field, as I return to the ordinary attending and my involvements with things, I discover that not only is the "world" of vision referred to me within experience overall, but that all the "objects" within that field of visual experience are *never* unsituated even within the field. Things or "objects" appear *only* as *essentially* situated *in* a field.

Within this preliminary glimpse certain "essential" structures or existential possibilities of visual experience are already anticipated. However, the purpose of this preliminary survey, which latently includes many phenomenological results, is to begin to model what in various modifications underlies the phenomenological notion of experiential intentionality. The illustration noted above can be diagrammed as shown (fig. 1).

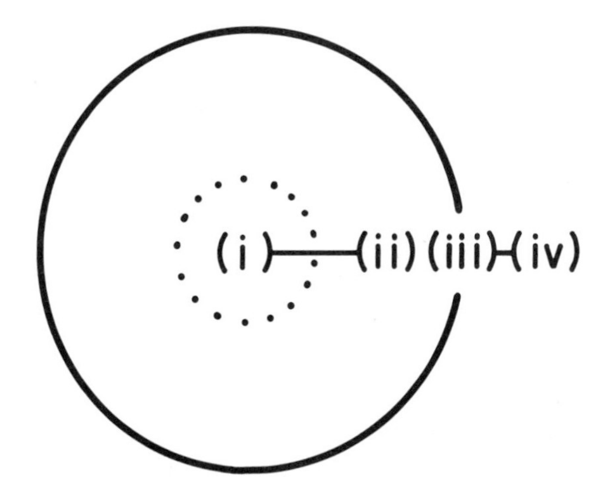

Noematically the appearances of the visual "world" in most ordinary experience display (i), a *focal core*, that which stands out before one, the central "object" or object range of the visual intentionality; (ii), the peripheral fringe, situated in relation to the core but never absent even if not explicitly noted; (ii) shades off to (iii), the *horizon*, which is the "border" or limit of the visual field and its "beyond."

Together (i), focus, and (ii), fringe, make up the totality of the visual field, the totality of explicit to implicit visual *presence*. The horizon (iii) is sensed as a limit to the "opening" which is the visual field, and this sense of limit is the first sense of horizon. But beyond the "edge" of the visual field *nothing* is given as present, the "beyond" of the horizon is an *absence*, or emptiness (iv). Thus horizon has two meanings from the outset.[8]

Within the purposeful naïveté of the phenomenological reduction this first step is one of attending to that which is experienced and to the "how" that which is experienced is presented. However, strictly copresent with the appearance is the *reflective awareness* of this experience as "my" experience. It is "I" who does the focusing; it is to "me" that the fringe appears as background, as the not-specifically-attended-to; and it is "I" who detects the strange boundedness and finitude of the visual field which raises the question of the World which lies "beyond" the finitude of "my" opening to the World.

Progressing now within this simplified framework, I return to my visual experience. I note that in ordinary experience there are certain patterns and resistances to the way in which these structures function. For example, no matter how hard I try, I cannot extend my horizon as limit. It remains at the "edge" of the visual field, and as I turn my head it "turns," too, but in such a way that it remains an absolute if vague "edge" while what is central also remains *before me*.

I also notice that ordinarily there remains a discernible *ratio* of the explicit to the implicit in relation to my attending acts. By exercising a series of variations I begin to find that there are, however, certain variable qualities to this ratio. The ratio of core-to-fringe may be exceedingly "narrow" in a "fine focus" of a visual act. I look at the tip of my pencil, taking in its grain, texture, leaden quality. But as I observe in a narrowly and finely focused mode the fringe "comes in", and that which is implicit and vaguely present, while still situating and surrounding the core, presents itself as covering most of the visual field.

But I can also "expand" the focus more "widely" and take in within central vision those faces before me in the classroom. Here the core area of clarity has expanded, and the fringe recedes farther towards the horizon. However, this variability seems to be limited in ordinary experience to a relatively variable ratio of central to peripheral core and fringe.

I push the variations more extremely and wonder if this ratio can be even further expanded. Can it, for example, become so "broad" that it stretches to the horizon? I reflect upon my experience and discover that there are exceptional instances which approximate even this possibility, but they are not "ordinary" in the sense of belonging to the usual daily activities of ongoing involvement with the environment.

I recall lying on my back in a sunny field in a state of youthful boredom. The world appeared to me as "flat," "all the same"; it presented itself as "indifferent." Phenomenologically, I was attending to nothing-in-particular, and the focal core itself receded toward a limit of disappearance in the *blank stare* of boredom. I shall call this a *field state*. Another variation in contrast to this is the state of ecstasy such as a first experience of a natural wonder. I remember the first time I came upon the Grand Canyon. It was at dusk when the whole panorama stretched before me with its blue and purple hazes, and in a brief moment of speechlessness *everything* seemed to be transformed. Again, while the ratio of core to fringe did not entirely disappear, here the focal core "expanded" with intensity toward the very horizons of the visual field as a panoramic whole.

Both the blank stare and the ecstatic vision reveal something of the heightened or depressed appearances of the phenomenon of the world. For in phenomenology *every mood* reveals something about the quality of an appearance. In the state of boredom the visual "world" lacks its normal sense of involvement "with me," whereas in the state of ecstasy the "whole world" leaps out "towards me" in its beauty and awesomeness. Only the viewer who has refused to recognize this and has subtly assumed or presupposed that a "neutral" state of observation is "normal" views such differences "abstractly" and fools himself into believing that the visual qualities of this range of experiences are the "same." But this is not to descriptively analyze visual experience; this is to transform it into an "abstract" seeing. Nor is it accidental that the preference for a "neutral" or "abstract" state is prefęrred as a *standard* of vision by the "Cartesian."

This first survey of one dimension of experience is intended to illustrate paradigmatically some of the ways in which a phenomenological "psychology" operates. In terms of the initial Human ⇄ World correlation model the relational arrows are *constant*. They are not lifted. But *within* the constancy there may be seen to be a variable and "floating" movement of weighted focus. Not only may we shift our intentional involvement toward the "objects" or noema at the World side of the relation and then shift back reflectively to the act-quality of the intention, but the full implications of the "flux" of the quality of the type of involvement may shift. In the field state of boredom the quality of the relation is quite different from that of an ordinary state of involvements, and likewise in an ecstatic state the quality of the relation is different again. In this sense every "mood" has epistemological significance for the phenomenologist.

But, at a higher level of consideration, the constancy of the relation of human being *in* the World is what bespeaks the phenomenological sense of one's being "immersed" in the "surrounding world," and at the same time within that sense of the world of human being, as Merleau-Ponty put it, the human is "already outside himself in the world."[9] Within the correlation of phenomenological experience overall it is a relation with the world that is known.

If this is so, there remains the implication that all forms of world-knowledge within phenomenology are relational; but likewise, all forms of self-knowledge are also relational. At one level there may often be found a kind of symmetry between world-knowledge and self-knowledge which is symptomatic. For example, in historical and cultural myths there is often a typological similarity between what is taken as a primary world reality and human-reality. Thus in archaic cultures the hunted bear or other totem of the tribe was also a brother who was considered to have a spirit not unlike that of humans. The quality of the surrounding world of the so-called animistic societies reflected back upon the human self-understanding in relation to that environment. If contemporary man thinks of this symmetry as merely a "primitive anthropomorphism," then the question might equally be raised about a similar symmetry in the contemporary world. A technological culture increasingly seems to view the world not only in "mechanistic" terms but humankind as "like" a machine, even if the latest variant is that of a highly complex and programmed computer. Here the anthropomorphism functions as strongly as ever, only the model of the relational other is changed (computer for animal).

A simple parable of learning of the self *through* or reflected in the world can be hypothesized in terms of an imaginative self-learning concerning the eye.

Imagine an odd and restricted case of a visual "world" with but one viewer. Here there are no mirrors and no others, those more ordinary "reflectors" by which we learn of ourselves; yet the viewer "sees" this "world." He takes note of just those features outlined in this initial model of a "visual" intentionality and, through reflection, asks what must be the nature of his strange "opening" to the "world." He notes that the visual field is bounded by the roundish horizon. Is his "opening" also roundish? He notes his ability to focus and to vary the focus upon the face of the "world." Does he have a "variable" opening? And so forth. Of course this need not be the way we actually learn of ourselves and the way in which the pupil of the eye functions. Ultimately the example hides more than it reveals, because without others and language it remains even doubtful that there could be a full self-reflectivity at all (a reduction of the world implies a reduction of the self!).

In the actual world, others and mirrors reflect us to ourselves, but the principle is still the same in that it is only by being "outside myself in the world" that I gain the reflective self-knowledge which I have. Moreover the "first" experience with a "mirror" is often quite curious and causes a sense of wonder. Jane Goodall's wild chimpanzees were amazed and often would quickly walk around the mirror which she placed before them to see who was "in" or behind it. And although visual mirrors are introduced too early for most of us to recall our wondering reaction to them, our surprised reaction to our voices reflected by the auditory "mirror" of the tape recorder is striking.

With this first visualist approximation of a model of phenomenology, now with perceptual experience in mind the first extension of the model may be the *auditory turn.* In what ways does or doesn't the model apply to the auditory dimension? The claim of a Husserlian phenomenology is that directed intentionality with its range of possibilities is an essential or invariant structure of experience.

However, with the question of an auditory turn there also arise serious preliminary questions which must be considered. What is involved with the examination of a "region" or dimension of experience? Here stands a crossroads at which phenomenology may make its own way or become confused by highly sedimented and accepted traditions concerning experience.

What is involved in a "reduction" *to* listening? Or is a "reduction" to listening even possible? The question may seem strange, exactly because we are accustomed by old habits of supposing and philosophical thought to seemingly do just that. What is more obvious than the five senses? This is precisely the danger point where the very first step of *epoché* could founder.

If there is anything to be drawn from previous work in phenomenology, particularly from work concerned with perception, the first result should be to understand that the primordial sense of experience is *global.* For Husserl this emphasis comes across repeatedly in his insistence that it is the same thing which presents itself in various profiles and in the various modes of experience. For Merleau-Ponty and for Heidegger the primordial experiences of being embodied or incarnate in a world are, if anything, even more strongly dependent upon the global character of primordial experience. In its ongoing and normal sense, experience in its first naïveté is not experienced as being constructed from parts. And as Merleau-Ponty has so clearly shown, even at the theoretical level a theory of perception is already a theory of the body and vice-versa.[10] In an existential phenomenology it is the body-as-experiencing, the embodied being, who is the noetic correlate of the world of things and others.

Yet the ease with which we assume a "reduction" to *a* sense remains as an easily taken-for-granted possibility which is sedimented in an old and particularly empiricist tradition. We "believe" that we can isolate one sense from the others; we "believe" that we "build up" or "synthesize" an object out of "sense data" or some other form of "sensory atom." These "beliefs" lie deeply imbedded in recent times with the "sense atomism" which infects even the sciences at their "metaphysical" level.

But a phenomenological "empiricism" inverts this understanding.

Its own scrutiny of experiential phenomena shows as foundational that at the first level the "synthesis" is what appears. Even a rather superficial reflection upon normal and ordinary ongoing experience would show that we have no conscious awareness of "processes" which gather data, and then which "build up" an object before us: the object "primitively" stands before us in all its diversity and richness and unity.

The reason such processes cannot be found lies within the metaphysical model long regarded as obvious in classical empiricism and in the even older traditions of metaphysical explanation. Ultimately sense data and primary qualities and a whole family of related unexperienced causes are ghosts which lie behind experience rather than lie *in* primordial experience. As an alternative view, phenomenology places in brackets precisely these "beliefs."

Thus the turn to a "pure" auditory experience becomes complicated with the rejection of a metaphysics of the five senses. But it is much easier to say that sensory atomism is placed out of play than to practice it, because *epoché* from the very first implies the double task of setting aside explanations and of isolating its selected region of description. Is there no sense in which the phenomenologist can speak about auditory experience "as such"? Surely the deaf person relates to the world in ways different from those of his neighbors with hearing, and the blind man relates to the world differently from his sighted peers. Does not the lack of a sense show something?

The answer must be a qualified "yes" but in terms dictated by the nature of the phenomenological inquiry. It must take its shape from its own methods and understandings. That the blind or deaf man experiences his problem with frustration, living as he must in a society of others who speak and who see would seem to indicate a sense of the "lack." Yet even the blind man experiences his perspective of the world as global, as a plenum.[11] We do not experience ultraviolet "light" as light, nor can we concretely even imagine what such an experience is like except as an empty supposing or by an analogy, but we do not experience this as a "lack." Were we suddenly to be plunged into a society of bees and there have to make our ways, we should then begin to appreciate quite dramatically this "lack" which lies beyond the threshold of our visual experience.

But the proximate way in which the auditory turn may be made lies at hand in the already exemplified distinctions of a ratio of focus-to-fringe and of the ratio of the explicit-to-the-implicit. *Within* global experience the model of a visualist intentionality applies in its own way. I can *focus* upon my listening and thus make the auditory dimension stand out. But it does so only relatively. I cannot isolate it from its situation, its embedment, its "background" of global experience. In this sense a "pure" auditory experience in phenomenology is impossible, but, as a focal dimension of global experience, a concentrated concern with listening is possible. Auditory experience can be thematized relatively, in relation to its contextual appearance within global experience. But just as no "pure" auditory experience can be found, neither could a "pure" auditory "world" be constructed. Were it so constructed it would remain an abstract world.

As an exercise in focal attention, the auditory dimension from the outset begins to display itself as a pervasive characteristic of bodily experience. Phenomenologically I do not merely hear with my *ears*, I *hear* with my whole body.[12] My ears are at best the *focal* organs of hearing. This may be detected quite dramatically in listening to loud rock music. The bass notes reverberate in my stomach, and even my feet "hear" the sound of the auditory orgy.

The deaf person—and most writers indicate that *total* deafness does not occur, since some hearing is by bone conduction with even highly deaf persons—has lost the use of his *focal* organs. He "hears" essentially differently than the normal listener. What are for normal listeners the fringe aspects of hearing, the feeling of the body of sounds which amplify the richness of focal hearing with the ears, are for the deaf person the "focus" itself. He is like a person with a central cataract obscuring his vision who perceives only the periphery of the visual field in terms of the proximate model described above. An approximation of this sense of "hearing" may be discerned in the following threshold phenomenon.

In Vermont while lying in bed at night my son often asked what the strange vibration of the earth was, until we noted that this vibration modulated into the clearly heard approach of a high-flying jet airplane some minutes after the first "felt" detection of its approach. Later we all recognized the transition of "felt" to "heard" sound which the jet displayed.

Sound permeates and penetrates my bodily being. It is implicated from the highest reaches of my intelligence which embodies itself in language to the most primitive needs of standing upright through the sense of balance which I indirectly know lies in the inner ear. Its bodily involvement comprises the range from soothing pleasure to the point of insanity in the continuum of possible sound in music and noise. Listening begins by being bodily global in its effects.

LISTENING AND VOICE

1. See Edmund Husserl, *The Crisis of European Science and Transcendental Phenomenology*, trans. David Carr (Evanston, Ill.: Northwestern University Press, 1970), pp. 191–257.
2. Maurice Merleau-Ponty, *Phenomenology of Perception*, trans. Colin Smith (London: Routledge and Kegan Paul, 1962), pp. viii–ix.
3. See Edmund Husserl, "Phenomenology," ed. Richard M. Zaner and Don Ihde, *Phenomenology and Existentialism* (New York: Capricorn Books, 1973), for one discussion of the reductions. Unfortunately the use of the term *reduction* has both a bad and a good sense in phenomenological usage. When referring to reductionism as contrasted to the reduction of mediate assumptions the sense is negative.
4. Edmund Husserl, *Cartesian Meditations*, trans. Dorion Cairns (The Hague: Martinus Nijhoff, 1960), p. 7.
5. *Ibid.*, p. 13.
6. P. F. Strawson, *Individuals* (London: Methuen, 1971). See chapter 2, pp. 59–86.
7. See Edmund Husserl, *Ideas*, trans. Boyce Gibson (New York: Collier Books, 1962), See chapter 9, pp. 235–59.
8. There is both a horizon to a particular thing with the implicit sense of absence and an absolute horizon to a field. Neither meaning of horizon should be confused with the ordinary signification of the distant line of earth and sky.
9. Merleau-Ponty, *Phenomenology of Perception*, p. xi.
10. Ibid., pp. 203.
11. Pamela Kay Haughawout, " 'I See' Said the Blind Man" (unpublished paper, 1969), p. 4.
12. This point is easily established physiologically. It has been repeatedly pointed out by Georg von Békésy in *Experiments in Hearing*, trans. E. G. Weaver (New York: McGraw-Hill, 1960), pp. 148 and 163–64.

PART TWO

DESCRIPTION

CHAPTER FOUR

THE AUDITORY DIMENSION

What is it to listen *phenomenologically*? It is more than an intense and concentrated attention to sound and listening, it is also to be aware in the process of the pervasiveness of certain "beliefs" which intrude into my attempt to listen "to the things themselves." Thus the first listenings inevitably are not yet fully existentialized but occur in the midst of preliminary approximations.

Listening begins with the ordinary, by proximately working its way into what is as yet unheard. In the process the gradual deconstruction of those beliefs which must be surpassed occurs. We suppose that there are significant contrasts between sight and sound; thus in the very midst of the implicit sensory atomism held in common belief we approximate abstractly what the differences might be between the dimensions of sight and of sound.[1] We "pair" these two dimensions comparatively. First we engage in a hypothetical and abstract mapping which could occur for ordinary experience with its inherent beliefs.

Supposing now two "distinct" dimensions within experience which are to be "paired," I attend to what is seen and heard to learn in what way these dimensions differ and compare, in what ways they diverge in their respective "shapes," and in what ways they "overlap."

I turn back, this time imaginatively, to my visual and auditory experience and practice a kind of free association upon approximate visual and auditory possibilities, possibilities not yet intensely examined, which float in a kind of playful revery.

Before me lies a box of paper clips. I fix them in the center of my vision. Their shape, shininess, and immobility are clear and distinct.

But as soon as I pair their appearance with the question of an auditory aspect I note that they are also *mute.* I speculatively reflect upon the history of philosophy with recollections of pages and pages devoted to the discussion of "material objects" with their various qualities and upon the "world" of tables, desks, and chairs which inhabit so many philosophers' attentions: *the realm of mute objects.* Are these then the implicit standard of a visualist metaphysics? For in relation to stable, mute objects present to the center of clear and distinct vision, the role of *predication* seems easy and most evident. The qualities adhere easily to these material objects.

A fly suddenly lands upon the wall next to the desk where the paper clips lie and begins to crawl up that wall. My attention is distracted and I swat at him. He quickly, almost too quickly for the eye, escapes and flies to I know not where. Here is a moving, active being upon the face of the visual "world." With the moving, active appearance of the fly a second level or grouping of objects displays itself. This being which is seen is active and is characterized by motion. Movement belongs to the verb. *He walks, he flies, he escapes.* These are not quite correctly properties but activities. Who are the "metaphysicians" of the fly? I recall speculatively those traditions of "process" and movement which would question the dominance of the stable, mute object, and which see in motion a picture of the world. The verb is affirmed over the predicate.

But the metaphysicians of muteness may reply by first noting that the moving being appears against the background of the immobile, that the fly is an appearance which is discontinuous, that motion is an occasional "addition" to the stratum of the immobile. The fly's flight is etched against stability, and the arrow of Zeno, if it may speed its way at all, must do so against the ultimate foundation of the stable background. Even motion may be "reduced" to predication as time is atomized.

But what of sound? The mute object stands "beyond" the horizon of sound. Silence is the horizon of sound, yet the mute object is silently *present.* Silence seems revealed at first through a visual category. But with the fly and the introduction of motion there is the presentation of a buzzing, and Zeno's arrow whizzes in spite of the paradox. Of both animate and inanimate beings, motion and sound, when paired, belong together. "Visualistically" sound "overlaps" with moving beings.

With sound a certain liveliness also makes its richer appearance. I walk into the Cathedral of Notre Dame in Paris for the first time. Its emptiness and high arching dark interior are awesome, but it bespeaks a certain monumentality. It is a ghostly reminder of a civilization long past, its muted walls echoing only the shuffle of countless tourist feet. Later I return, and a high mass is being sung: suddenly the mute walls echo and reecho and the singing fills the cathedral. Its soul has momentarily returned, and the mute testimony of the past has once again returned to live in the moment of the ritual. Here the paired "regions" of sight and sound "synthesize" in dramatic richness.

But with the "overlapping" of sight and sound there remains the "excess" of sight over sound in the realm of the mute object. Is there a comparable area where listening "exceeds" seeing, an area beyond the "overlapping" just noted where sight may not enter, and which, like silence to sound, offers a clue to the horizon of vision?

I walk along a dark country path, barely able to make out the vague outlines of the way. Groping now, I am keenly aware of every sound. Suddenly I hear the screech of an owl, seemingly amplified by the darkness, and for a moment a shock traverses my body. But I cannot see the bird as it stalks its nocturnal prey. I become more aware of sound in the dark, and it makes its presence more dramatic when I cannot see.

But night is not the horizon of sight, nor Dionysius the limit of Apollo. I stand alone on a hilltop in the light of day, surveying the landscape below in a windstorm. I hear its howling and feel its chill, but I cannot see its contorted writhing though it surrounds me with its invisible presence. No matter now hard I look, I cannot see the wind, *the invisible is the horizon of sight*. An inquiry into the auditory is also an inquiry into the invisible. Listening makes the invisible *present* in a way similar to the presence of the mute in vision.

What metaphysics belong to listening, to the invisible? Is it also that of Heraclitus, the first to raise a preference for vision, but who also says, "Listening not to me but to the Logos, it is wise to acknowledge that all things are one."[2] Is such a philosophy possible beyond the realm of mute objects? Or can such a philosophy find a way to give voice even to muteness? The invisibility of the wind is indicative. What is the wind? It belongs, with motion, to the realm of verb. The wind is "seen" in its *effects*, less than a verb, its visible being is what it has done in passing by.

Is anything revealed through such a playful association? At a first approximation it seems that it is possible to map two "regions" which do not coincide, but which in comparison may be discerned to have differing boundaries and horizons.

In the "region" of sight there is a visual field which may be characterized now as "surrounded" by its open horizon which limits vision, and which remains "unseen." Such a field can be diagrammed (fig. 2).

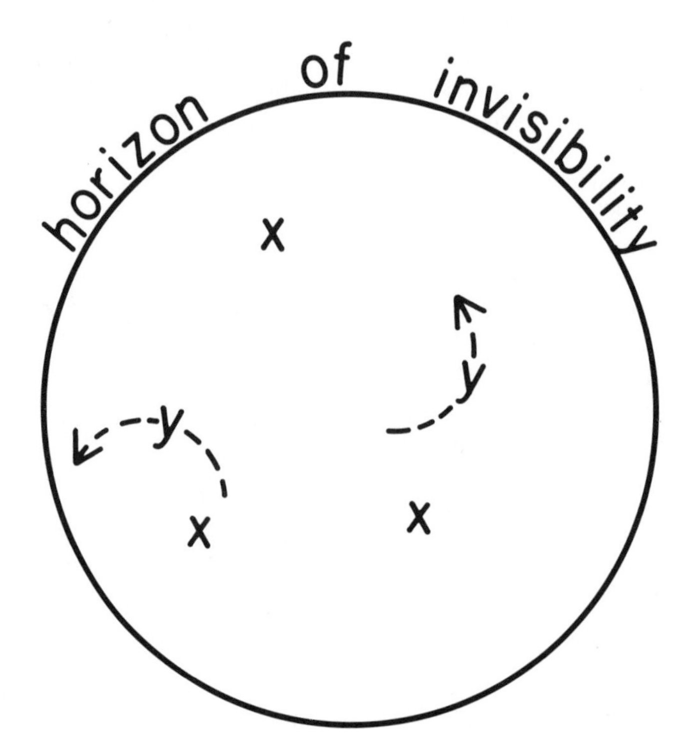

Here, where the enclosed circle is the present visual field, within this presence there will be a vast totality of entities which can be experienced. And although these entities display themselves with great complexity, within the abstraction of the approximation we note only that some are stable (x) and usually mute in ordinary experience, and that some (--y→) move, often "accompanied" by sounds. Beyond the actually seen field of presence lies a horizon designated now as a horizon of invisibility.

A similar diagram can be offered for a "region" of sound presences (fig. 3).

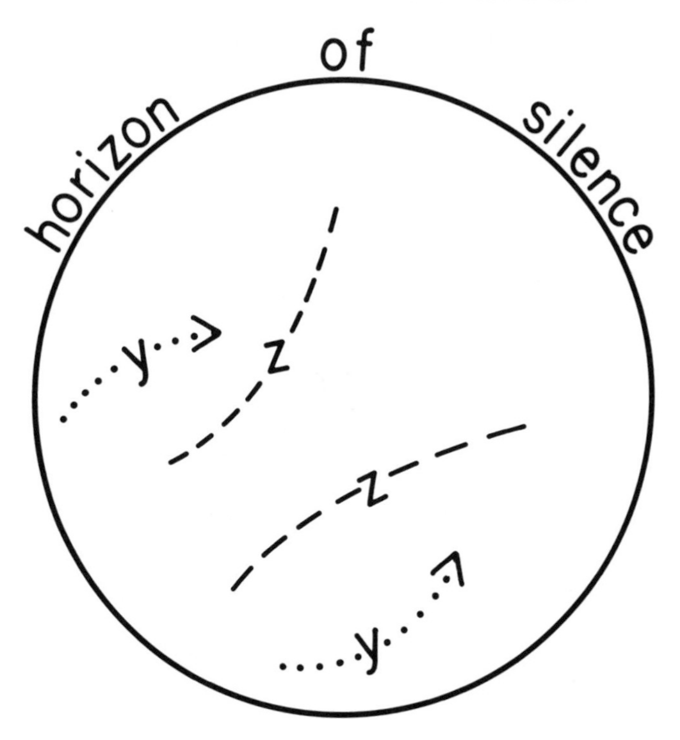

Although once we move beyond this approximation, the "shape" of the auditory field will need to be qualified. Within the limits of the first approximation we note that the auditory field contains a series of auditory presences which do not, however, perfectly overlap those of the visual field. There are sounds which "accompany" moving objects or beings (--y--), but there are some for which no visible presence may be found (--z--). Insofar as all sounds are also "events," all the sounds are, within the first approximation, likely to be considered as "moving." Again, there is also a horizon, characterized by the pairing as a horizon of silence which "surrounds" the field of auditory presence.

It is also possible to relate, within the first approximation, the two "regions" and discern that there are some overlapping and some non-overlapping features of each "region." Such a "difference" may be diagrammed (fig. 4).

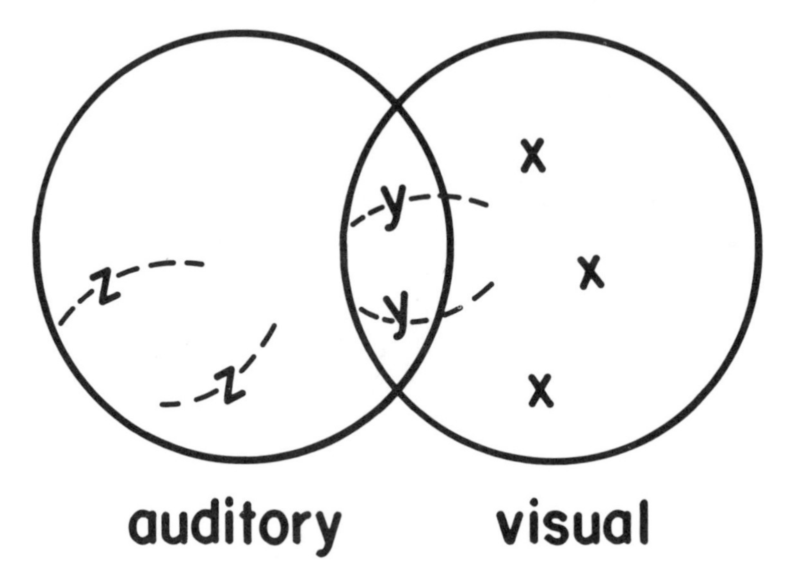

In this diagram of the overlapping and nonoverlapping "regions" of sight and sound we note that what may be taken as horizonal (or absent) for one "region" is taken as a presence for the other.

Thus while the area of mute objects (x) seems to be closed to the auditory experience as these objects lie in silence, so within auditory experience the invisible sounds (--z--) are present to the ear but absent to the eye. There are also some presences which are "synthesized" (--y--) or present to both "senses" or "regions."

This pairing when returned to the revery concerning the associated "metaphysics" of the "senses" once more reveals a way in which the traditions of dominant visualism show themselves. If we suppose that any metaphysics of worth must be one which is at least comprehensive, then a total visualist metaphysics must find a way to account for and to include in its description of the world all those invisible events which at this level seem to lie beyond the reach of the visible horizon, but which are nevertheless present within experience.

This may be done in several ways. First, one can create some hermeneutic device which, continuing the approximation of the "regions," *functionally* makes the invisible visible. This implies some "translation" of one "region" into the terms of the favored "region." Such is one secret of the applied metaphysics often found in the sciences of sound. Physically, sound is considered a wave phenomenon. Its wave characteristics are then "translated" into various visual forms through instruments, which are the extended embodiments of the scientific enterprise. Voice patterns are "translated" into visual patterns on oscillographs; sound reverberations are mapped with Moire patterns; even echo-location in its practical applications is made a matter of seeing what is on the radar screen: the making or "translating" of the invisible into the visible is a standard route for understanding a physics of sound.

In the case of the sciences of sound this translation allows sound to be measured, and measurement is predominantly a matter of spatializing qualities into visible quantities. But in ordinary experience there is often thought to be a similar role for sound. Sounds are frequently thought of as anticipatory clues for ultimate visual fulfillments. The most ordinary of such occurrences are noted in locating unseen entities.

The bird watcher in the woods often first hears his bird, then he seeks it and fixes it in the sight of his binoculars. The person hanging a picture knows where to look for the dropped tack from the sound it made as it rolled under the piano. And although not all noises yield a visual presence—for example the extreme case of radio astronomy may yield the presence of an unsuspected "dark" star which may never be seen—the familiar movement from sound to sight may be discerned.

The movement from that which is heard (and unseen) to that which is seen raises the question of its counterpart. Does each event of the visible world offer the occasion, even ultimately from a sounding presence of mute objects, for silence to have a voice? Do all things, when *fully* experienced, also sound forth?

In ordinary experience this direction is also taken. The bird watcher may be an appreciative bird listener. He awaits quietly in the hopes that the winter wren will sing his long and complicated "Mozart" song. But only in more recent times has this countermovement become conspicuous. The amplified listening which now reveals the noise of lowly ant societies gives voice to the previously silent. Physically even molecules sound, and the human ear comes to a threshold of hearing almost to the point of hearing what would be incessant noise.[3]

1. A phenomenological warning must be issued here. There is a strict difference between empty supposing and what is intuitionally fulfilled. Thus the exercise at this point is not strictly phenomenological but proceeds toward strict phenomenology by approximations.
2. Philip Wheelwright, *The Presocratics* (New York: Odyssey Press, 1966), p. 79.
3. Georg von Békésy, *Sensory Inhibition* (Princeton: Princeton University Press, 1967), p. 9.

CHAPTER FIVE

THE SHAPES OF SOUND

The approximation which opened a difference between sight and sound ended in a questioning of the import of that difference. If a movement is possible which gives visibility to the unseen, and a countermovement which gives voice to the mute is possible, a closer listening to the auditory dimension itself is called for. The time has come when listening must begin to be reflective. I begin to take note of my listening, and I first notice a certain incessant field of sound which strikes me as a constant "flux" marked by an obvious and dramatic "temporality."

I begin to catalogue my auditory experience within a given moment of time, and I note that within only a few moments a series of sound-events have occurred. There is the sound of the vacuum cleaner on the floor below; just then there was the pounding of the construction worker next door; the rustle of leaves is heard momentarily; and, if I am more attentive to less obvious sounds, there is the buzz of the fluorescent light and the hum of the heating system. But I also conclude, perhaps too easily and too quickly, that the auditory world is one of "flux" and that it is primarily *temporal.*

I close my eyes and note that one sound follows another, that a single sound "exists" for a moment and "passes away," and that there is an "inconstancy" to this "region" in which the surging of time is dramatically present. This intimacy of temporality with the auditory experience forms a central tradition concerning sound and may be found recorded by philosophers as diverse in points of view as Søren Kierkegaard, Edmund Husserl, and P. F. Strawson. What I "discover" first is already known and sedimented as knowledge.

In meditating upon music and language as sensuous media Kierkegaard writes, "The most abstract idea conceivable is sensuous genius. But in what medium is this idea expressible? Solely in music . . . it is an energy, a storm, impatience, passion, and so on, in all their lyrical quality."[1] But also noting the auditory dimension of language he states, "Language addresses itself to the ear. No other medium does this. The ear is the most spiritually determined of the senses . . . aside from language, music is the only medium that addresses itself to the ear."[2]

Language and music, auditory phenomena, are understood by Kierkegaard to be dominantly temporal in their actual form. "Language has time as its element; all other media have space as their element. Music is the only other one that takes place in time."[3] This *positive* relation of sound to time is what contextually appears as "first" in a reflective listening. It is also maintained within phenomenology in the use of auditory material in Husserl's *Phenomenology of Internal Time Consciousness.* Not only are his usual visual examples often and even dominantly replaced in these lectures by auditory ones, but even the use of metaphorical and descriptive language begins to take on an auditory tone. "The bird changes its place; it flies. In every situation the echo of earlier appearances clings to it (i.e., to its appearance). Every phase of this echo, however, fades while the bird flies farther on. Thus a series of '*reverberations*' pertains to every subsequent phase, and we do not have a simple series of successive phases."[4] With Kierkegaard, Husserl takes note of the overwhelming intimacy of sound and time.

However, where this is "traditional" concerning sound, and where this strong tie cannot be overlooked in any analysis of auditory experience, there is often either implicitly or explicitly a negative claim that listening is either therefore "weak" spatially or, most extremely, that *sound lacks spatiality entirely.* This negative claim is most blatant in precisely that tradition which most clearly "atomizes" the senses and reduces them to their lowest forms: the empiricist tradition. This view, rapidly losing ground in the biological and physical sciences, sometimes affirms that "a spatial order is connate only for the optical, tactile, and kinesthetic spheres; while for the other senses, mere complexes of feelings with spatial features are admitted."[5]

Such a view is explicit in Strawson's *Individuals.* Strawson, clearly defending a "metaphysics of objects" in an Aristotelian vein which continues the visualism of vision and objectification, considers a "No-Space world" which he finds a conceptual possibility of a reduced world of "pure" sound.

> The fact is that where sense experience is not only auditory in character, but also at least tactual and kinesthetic as well—we can sometimes *assign* spatial predicates on the strength of hearing alone. But from this fact it does not follow that where this experience is supposed to be exclusively auditory in character, there would be any place for spatial concepts at all. *I think it is obvious* that there would be no such place.[6]

What is "obvious" is that a tradition is here being taken for granted with disregard for the contemporary discoveries of very complex spatial attributes to auditory experience. Directionality and location, particularly advanced in such animals as porpoises and bats but not lacking in humans, have shown the degree to which echo-location is a very precise spatial sense. The bat's ability to "focus" a "ray" of sound such that he may discern the difference between a twig and the moth he is after is now well known. But such auditory abilities have long been encased in precisely the tradition which denies spatiality to listening and for decades and even centuries prevented the scientist from believing that it was indeed a capacity for sound and listening.

Although experiments with bats as early as those of Lazzaro Spallanzani in 1799 led him to ask "whether their ears rather than their eyes serve to guide them in flight." The already established prejudices of the ancients caused even Spallanzani to doubt his findings. Even the suggestion that hearing could detect and localize objects "in space" was vigorously attacked by eminent figures such as Georges Cuvier and George Montagu. It was not until 1912 that the suggestion that hearing was "spatializable" reopened the question which has led to contemporary knowledge concerning echo-location in a whole series of animals, and which today may lead to the development of amplification devices by which the blind may extend their often already acute hearing.

It is precisely because of the very "obviousness" *not* of experience, but of the traditions concerning experience that there is reason to postpone what is "first" in the turn to the auditory dimension. Without denying the intimacy of sound and time and without denying the richness of the auditory in relation to temporality, a strategy which begins in approximations is one which must move with extreme care so as not to overlook or fail to hear what also may be shown in the seemingly weaker capacities of auditory experience. Thus as the move into phenomenology proper is made, it is with the *spatiality* of sound that description may begin. Within a spirit of gradual approximations the "weakest" possibilities of sound are to be explored before the "strongest" possibilities.

However, there are several initial qualifications which must be held in mind in beginning phenomenological description in this way. First, the movement from the more abstract approximation which began in the midst of sensory atomism is one which not only increasingly accelerates away from that division of the senses, but one which begins in making thematic what will be called here the first existential level of experience, the level of "greatest naïveté." For despite the extreme technicality of Husserl's discussion of identity from *Logical Investigations*, the outcome is one which reaffirms the primacy of the *thing* in naïve or first existential experience. It is to *things* that we attend in naïve and ordinary experience once we have set aside our layers of beliefs regarding how those things "should" present themselves.

The same applies to auditory experience. Sounds are "first" experienced as sounds *of* things. That was the sound *of* the jackhammer with all its irritating intrusion. There, it's Eric calling Leslie now. That was definitely a truck which went by rather than a car. This ease which we take for granted and by which we "identify" things by sound is part of our ongoing ordinary experience. This common ability of listening contains within it an extraordinary richness of distinction and the capacity to discern minute differences of auditory texture, and by it we know to what and often to where it is that our listening refers.

Often we find extraordinary examples of these capacities in the *musical ear*. Beethoven, for example, had such a rich and extraordinary auditory ability, both perceptual and imaginative, that he could compose and imaginatively hear a whole symphony in his head and specifically discouraged anyone from using the piano to demonstrate passages, because the piano was much poorer than the whole symphony in his head. But this musical, perceptual memory, though not equally acute, is not rare among accomplished musicians.

Such musical feats are also potentially misleading, because there is also a tradition, echoed above by Kierkegaard, that music is "abstract." Even phenomenologists have been misled to take the musical experience as one which is disembodied and "separated from its source" as a kind of "pure" auditory experience.[7]

In daily concerns such abstract listening is at least unusual, yet its feats of discernment are highly discriminating. On walking along a village street in Llangefni, Wales, my son pointed out a thrush busily banging a snail against the sidewalk. This act soon successfully produced a tasty meal even without benefit of garlic and butter. Several weeks later I was awakened in our house in London to the early morning unmistakable cracking of the snail shell coming through the curtained window. I drew the curtains to show my wife this occurrence which was new to her, but the "identification" had been quite "obvious" to me by the single sound of the cracking snail shell.

Such identifications and discriminations of minute auditory differences are not as yet "spatial." But having made a turn of attention to the first naïve existential level of experience where sounds are the sounds of things, the spatial aspects of that experience may begin to show themselves. In searching out the spatiality of sound the cautions previously noted take specific form here. First, auditory spatiality must be allowed to "present itself" as it "appears" within this level of experience. Negatively, a predefinition of spatiality such that it is prejudged "visualistically" must be suspended.

Second, affirming the phenomenological sense of the global character of primal experience, it is necessary to replace the division of the senses with the notion of a *relative focus* upon a dimension of global experience such that it is noted only against the omnipresence of the globality. Thus a "pure" experience is eliminated and made impossible. Primitively things are always already found "synthesized" in naïve existential experience. The move to a focus-fringe interpretation of global experience thus safeguards the tendency towards disembodiment which tempts all "Cartesian" types of philosophy, and which mixes, in spite of itself, perceptual and emptily suppositional terms.

Third, as a first phenomenological approximation in contrast to the approximation in the midst of sensory atomism, it should be noted that even the division of space and time are not, strictly speaking, primitive experiential significations. Existentially there is a concrete space-time which is also a signification of naïve experience in its thematized appearance.

With this second approximation, the entry into the "weakness" of the auditory dimension, phenomenological description proper begins. The provisional character of the sounds of things in ordinary experience should not be considered a final but a first level of the phenomenological experiential analysis.

Shapes, Surfaces, and Interiors

At the experiential level where sounds are heard as the sounds of things it is ordinarily possible to distinguish certain *shape-aspects* of those things. The following variations begin in what for human hearing is admittedly one of the weakest existential possibilities of listening. Nor do I claim that every sound gives a shape-aspect (but neither does every sighting give a shape-aspect in the ordinary sense). At first such an observation seems outrageous: *we hear shapes.*

The shape-aspects which are heard, however, must be strictly located in terms of their auditorily proper presentation and not predetermined or prelimited by an already "visualist" notion of shape. The shape-aspects which are heard are "weaker" in their spatial sense than the full outline shape of a thing which is ordinarily given all at once to vision. But a "weakness" is not necessarily a total absence, for in this "weakness" there remains an important, if primitive, spatiality for hearing.

Children sometimes play an auditory game. Someone puts an object in a box and then shakes and rolls the box, asking the child what is inside. If, more specifically, the question is directed toward shapes, the observer soon finds that it takes little time to identify simple shapes and often the object by its sound. For example, if one of the objects is a marble and the other a die (of a pair of dice), and the box is rolled, the identification is virtually immediate. The difference of shape has been *heard*, and the shape-aspect has been auditorily discriminated.

But the flood of likely objections to such an observation, however experientially concrete such examples are, threatens to overwhelm the listener. For in spite of the hermeneutic rules of *epoché* which attempt to put out of play both "sensory atomism" and its preferred "visualism," it threatens to return at each stage of analysis. It is precisely the recalcitrance of such beliefs which makes the act of auditory discernment "difficult to believe" in spite of one's ears.

The point here is not to enter into an interminable and difficult argument but to let the things show or sound themselves. For involved in the "weakness" of auditory spatiality there are a number of factors which allow that "weakness" to be missed if one is not careful in listening. What is amazing, however, is what appears spontaneously in the simple variation. The very first time I played this game with my son I had placed a ball-point pen in a box without his seeing it and rolled it back and forth. I asked him what shape it was. His answer was, "It's like a fifty-pence shape, you know, on its sides, only it's longer." A fifty-pence coin has seven sides, the ball-point pen had six, and it was, in his parlance, "longer."

Nor is the shape-aspect all that is given in the richness of simple auditory presentations. If the game is allowed to continue so that one learns to hear things in an analogue to the heightened hearing of the blind man's more precise listening to the world, a quickly growing sophistication occurs. A ball-point pen gives a quite different auditory presentation with its plastic click from that of a wooden rod. A rubber ball is as auditorily distinct from a billiard ball as it is visually distinct. The very texture and composition as well as the shape-aspect is presented in the complex richness of the event.

It is often this learning itself which offers itself as suspicious to the "sensory atomist" whose notion of a built-up or constructed knowledge also infects his understanding of learning. Phenomenologically there is a great distinction between *constructing* something and its *constitution.* In constitution the learning that occurs is a learning which becomes aware of what there is to be seen or heard. There is the usual inversion called for in *epoché* here. As Merleau-Ponty remarked, "*Learning* is *In der Welt Sein*, and not at all that *In der Welt Sein* is *learning.*"[8] This difference may be illustrated by two vastly different ways in which perceptual experience is employed in the empirical sciences.

In some psychology many of the experiments are deliberately designed to first disrupt all previous "learning" by radically altering its context. To view a white sheet of paper under blue lighting through a darkened tube which cuts off the normal context and field significance of the experience is to radically alter ordinary experience. But the learning which is tacit in ordinary experience is then further cut off by allowing the experience to continue for only an atom of time, thus preventing any adjustment. In this way the experiment is set up so that it often cannot help but circularly reenforce the "abstraction" of the "sensory atomist's" view of perception which begins with the "abstraction" of "sense data" or similar "stimuli." The experiment constructs the condition for the preformed conclusion and interprets what it finds as a *primitive* of experience.

Yet the always-present learning through which perceptions are incarnate functions here as well; only in this case it operates tacitly in the situation of the observer, the psychologist. Were he to be replaced by another observer as experimentally "naïve" as his subject, in all likelihood there would be little purpose to or knowledge resulting from the experiment. The observer would still have to enter the scene even if now his taken-for-granted judgments are removed one step further.

There is a sense in which the role of constitution proposed by the phenomenologist is implicitly recognized in the natural sciences. An ornithologist friend once described to me the pains he had to go through to get his first-year students to even produce a recognizable description of bird behavior. He would lapse into laughs when a report returned stating, "The bird *sat* on the fence, then it hopped and sat again." For in his parlance a bird not only does not sit, it perches; but the student in this case had not yet learned to see. At first the learner does not recognize the differences between the various species of warblers, which are often confusing anyway, but after long and careful learning he then wonders why he could not at first recognize what is now so obvious.

But the learning does not *construct* what is to be seen, it *constitutes* it in terms of its meaning. What is to be seen is *there*, and anyone entering this region of knowledge may see the distinctive marks which differentiate one warbler from the other. Once the distinctions have been learned, the previous lack of awareness and lack of discrimination is seen not as a fault of the "object" but of the inadequacy of our own prior observation.

This problem is partially due to our frequent failure to discern the space-aspects of auditory experience. We have not learned to listen for shapes. The whole of our interpretation in its traditional form runs from it, and only in the dire situation of being forced to listen for shapes, such as the advent of blindness, do ordinary men attend specifically to the shape-aspects of sound.

But even here there is a complication which arises from the global or plenary quality of primary experience. For the blind or deaf person experiences his "world" as a unity and his experience as a plenum. His sense of lack is conveyed by the transcendence of language, and he even becomes quite adept at 'verbalisms', the ability to define things through words, although he may not recognize them when they are presented to him. One blind person describes this sense.

> Those who see are related to me through some unknown sense which completely envelops me from a distance, follows me, goes through me, and, from the time I get up to the time I go to bed, holds me in some way in subjection to it.[9]

It is here that the "sensory atomist" finds so much "evidence" for his constructionist view of the world. It is well known that many, indeed most, persons who are blind cannot visually recognize certain objects presented to them until they feel these objects. But there is another possible interpretation of such "evidence." It is not that the object is built up, but that the learning which goes on in all experience must go on here, too. The radically new experience of seeing, when a blind person gains sight through a medical procedure, is revealing. His *first* sight, when reported, often turns out to be precisely "like" those first impressions reported in the first turn to reflective *listening*. He is impressed by what we might call the *flux* and *flow*, the implicit temporality of the new dimension to his experience. J. M. Heaton reports that when the blind are given sight, "at first colours are not localized in space and are seen in much the same way as we smell odours."[10] Odors, sounds, tastes, upon *first* note, appear not as fixed, but as a flux and flow. The first look is a stage of experience, not something which belongs isolated within one "sense."

This learning is often painful. For the patient it is not a mere addition to his experience but a transformation of the whole previous shape of the plenum of his experience.

> The chief difficulty experienced by these patients is due to the general reorganization of their existence that is required, for the whole structure of their world is altered and its centre is displaced from touch to vision; and not only perception but language and behavior also have to be reoriented.[11]

It is not, however, that upon being given sight spatiality is first discovered. It is *re-constituted.* A subtle example of this was given to me by a student trained in phenomenology who had been blind, but who, through treatment, gained limited sight. She noted that one quite detectable difference in her lived spatial organization when given sight was a gradual displacement of a previously more omnidirectional orientation and spatial awareness to a much more focused *forward* orientation. Although she noted that even when blind there was a slight "preference" for a forward directed awareness, this became much more pronounced with the gaining of vision.[12] Again, as will become more apparent as the spatial significations of the auditory dimension become more pronounced, the relative omnidirectionality of awareness and orientation is "closer" to the space-sense of sound than that of vision.

In a gradual clarification of the distinctive spatial sense of auditory experience, the first discrimination of shape-aspects heard in such spontaneous experiences as that of the game of placing an object in a box becomes more precise when attention is paid not only to the presence of the spatial aspect, but to how it is given in perception. Reverting to the pairing of sight and sound, this factor becomes easier to locate.

I turn to my visual and auditory experiences. I note now that in both dimensions there is a multiplicity of phenomena, but I also note that these do not always overlap. I see before me the picture of the sailboat, the note concerning last night's sherry party, a postcard from Japan. But I hear the cement mixer, the bird song, and the traffic in the street.

Next, I note that it seems at first that every stable thing before me visually presents a spatial signification which is, moreover, given-all-at-once. Each object has at least an *outline shape*, and this shape in the objects mentioned is discerned immediately. But of the sounds I do not seem to get shapes, certainly not outline shapes and certainly not all-at-once.

In comparing this nonoverlapping of shape in sight and sound in terms of the question of how shape-aspects are given, I soon find that the question of time is involved as well. The all-at-onceness to the outline shape before me is a matter of *temporal instantaneousness* or of *simultaneity*. But when I return to those experiences which give me shape-aspects I find that the one given is not a matter of instantaneousness but of a *sequential* or *durational* presentation. If the ball is dropped and does not bounce, I may not get more than a "contact point" as a vague and extremely "narrow" signification. But if the ball is rolled for several instants, if the rolling endures through a time span which is quite short, I get a sense of its shape as an *edge-shape*. This shape is presented not in terms of temporal instantaneousness but in terms of temporal duration. In both cases there is a need for some "time," as even visually the object presented in too small an atom of time remains equally spatially indiscernible. But there is a difference of need here in which the temporal duration for the discrimination of an edge-shape by sound must be relatively greater. Here again a clue seems to emerge as to why tradition has maintained the asymmetries of "spatial poverty" for sound and "temporal richness" for sound in comparison to the "spatial richness" and "temporal poverty" for sight.

But this comparative variation bespeaks only one, albeit important, variation in relation to spatial significations, and with it the sedimentation of the dominance of the mute object for spatial significations remains. Further variations, however, tend to diminish the asymmetries to a degree. If I return to the pairing of sight and sound and introduce the (rapidly) moving thing into experience, a difference occurs. The arrow, the drop of water, the stone which appear before me falling or flying at certain speeds do not show themselves as clear and distinct shapes. They present themselves as "vague" shapes which reveal themselves only when the motion stops. (In some cases this can happen if the field is large enough and the speed slow enough for me to fix my eyes upon an object as it moves.) Some form of *fixing* is required to determine the clarity and distinction of the outline shape. Once again the stable and mute object returns as the hidden norm of visualist space significance.

Yet the "weak" or "vague" shape-aspect of the moving object is closer to the many shape-aspects which auditory experience yields in its constant flux. A duration is needed to discriminate shape in this constant motion. Thus if "extended," temporal duration which persists in the flux and motion of sound in time is what appears as the main presentational mode of heard shape-aspects. The much shorter and more "instant" norm of visual stability allows duration to be either overlooked or forgotten and thus apparently to be less important in the visual discrimination of spatial significance.

An edge-shape is "less" than the outline shape, but it is a shape-aspect nonetheless. It is as if the ear had to gradually gain this shape in its durational attention. It is from such temporal considerations that "linear" time metaphors may arise. In this respect auditory shapes seem on one side to be closer to tactile shape discriminations. The blind Indian who concludes that the elephant is like a snake, and who argues with another who thinks the elephant is like a rough wall, is not wrong but inadequate in his "observations" concerning the shape of the elephant. Were he a rigorous feeler of elephants he would not be satisfied with instant apodicity but would withhold his conclusion until he had covered the whole surface of the elephant. So, with listening for shape-aspects it often takes repeated and prolonged listenings until the fullness of the shape appears. This serves no useful purpose in daily affairs when a mere glance will do the same in less time. Thus we fail to hear what may be heard and pass over an existential possibility of listening.

A third variation shows that there is even less absolute difference between sight and sound when size is taken into account. The edge-shape is usually admittedly quite "small." The marble rolling in contact with the box or the die striking the box presents only a small aspect of itself. But visually there is a reversion to a sequential discrimination, too, if the thing is immense. If one stands below the skyscraper, it is unlikely that he will take in the whole at once. He allows his gaze to follow the outline of the building, and the gaze in relation to the vastness becomes a sequential following of the outline-shape. The all-at-onceness does again become possible if distance is increased, as, for example, when I see the whole skyscraper from above while stuck in a traffic pattern in an airplane above Manhattan. Again the comparative reign of the now "middle-sized" stable and mute object returns, and the comparative "weakness" and difficulty of auditory shape discrimination returns; but only now it is understood as a matter of relative distancing in space-time. It remains the case that the shape-aspect which is discerned auditorily in its "weakest" possibility is a spatial signification which is limited to a degree within the dimension of hearing.

There is another factor of the hearing of shapes which reveals itself in the "weakness" of hearing the shape of the thing: one which raises the question of *how* the thing is *voiced*. The mute object does not reveal its own voice, it must be given a voice. In the examples listed, for the most part, a voice is given to the object by some other object. One thing is struck by another, one surface contacts another, and in the encounter a voice is given to the thing.

There is clearly a complication in this giving of voice, for there is not one voice, but *two*. I hear not one voice, but at least two in a "*duet*" of things. I hear not only the round shape-aspect of the billiard ball rolling on the table, I also hear the hardness of the table. The "same" roundness is heard when I roll the billiard ball on its felt-covered table, but now I also hear the different texture of the billiard table. True, just as in listening to an actually sung vocal duet, I can focus auditorily upon either the tenor or the baritone; but my focal capacity does not blot out the second voice, it merely allows it to recede into a relative background. Thus in listening to the duet of things which lend each other a voice, I also must learn to hear what each offers in the presence of the other. The way in which the mute things gain or are given voices in my traffic with the world is an essential factor in all spatial signification in sound. The voices of things call for further attention.

Although only a massive shift in perspective and understanding will ultimately allow the fullness of auditory spatial significations to emerge, the movement from weaker to stronger possibilities of listening is one which increases our familiarity with such significations. Less strange than the notion of hearing shapes, *we also hear surfaces*. This auditory experience is involved with our ordinary experiences of things.

Who does not recognize the surface in the sound of chalk scratching? I hear footsteps in the hallway. (I can tell if it is Leslie in her heels or Eric in his tennis shoes,) or, when the walker steps on the tile its surface produces a characteristic clacking sound of hard heels. Then, the moment the person first steps into the living room the clacking changes to the dull thudding sound of footsteps on the rug.

Surfaces, which are more familiar to us than shapes, must also be heard in terms of a voice being given the things. Just as in the discernment of shape-aspects (and shape-aspects may grade off into surface significations) there is usually a duet of voices in the auditory presentation. Furthermore, there is often more than a surface signification, a signification which grades off at the upper end into an anticipation of hearing interiors. I hear the textural and compositional character of the thing and distinguish easily between the sound of a bell and that of a stick hitting pavement.

Unaccustomed as we are to the language of hearing shapes and surfaces, we may remain unaware of the full possibilities of listening. But the paradigm of acute listening given in the auditory abilities of the blind man often provides clues for subtle possibilities of the ordinarily sighted listener as well. The blind man through his cane embodies his experience through a feeling and a hearing of the world. As Merleau-Ponty has pointed out, he *feels* the walk at the end of his cane. The grass and the sidewalk reveal their surfaces and textures to him *at the end of the cane.* At the same time his tapping which strikes those surfaces gives him an auditory *surface-aspect.* The concrete sidewalk sounds differently than the boardwalk, and in his hearing he knows he has reached such and such a place on his familiar journey.

To be sure, the surfaces heard by the blind man or the ordinary listener are restricted surfaces. They lack the *expanse* which vision with its secret "Cartesian" prejudice for "extension" presents, because the auditory surface is the revelation of an often small region rather than the spreading forth of a vista. But within its narrowness a surface is heard.

But striking a surface and thereby getting a duet of the surface aspects of two things is not the only way in which the mute object is given voice, nor is it the only way in which sound reveals surfaces. For the blind man's tapping also gives an often slight but nevertheless detectable voice to things in an *echo. With the experience of echo, auditory space is opened up.* With echo the sense of distance as well as surface is present. And again surface significations anticipate the hearing of interiors. Nor, in the phenomenon of echo, is the lurking temporality of sound far away. The space of sound is "in" its timefulness.

The depth of the well reveals its auditory distance to me as I call into its mouth. And the mountains and canyons reveal their distances to me auditorily as my voice re-sounds in the time which belongs so essentially to all auditory spatial significations. But these distances are still "poorer" than those of sight, though distances nonetheless. This relativity of "poverty" to "wealth" is apparent in the occasional *syncopation* of the visual and auditory appearances of the thing. Such a common experience today may be located in the visual and auditory presentations of a high-flying jet airplane. When I hear the jet I may locate its direction quite accurately by its sound, but when I look I find no jet-plane. The sound of the jet trails behind its visual appearance and, by now accustomed to this syncopation, I learn to follow the sound and then look ahead to find the visual presence of the jet.

But as I come to smaller distances the syncopation lessens, and the sight and sounds converge so that ordinarily the sight and sound of the things seem to synthesize in the same place. Yet with careful attention as I stand in the park and listen to the automobiles and trucks rush past, I find that even here there is a slight trailing effect. I close my eyes and follow the sound which, upon opening my eyes, I find only slightly trails the source as seen. Soon I can detect this trailing with my eyes open. Again in this distance the temporality of sound is implicated.

This often unpracticed and unnoticed form of human echo-location which is spatially significant may also be heightened. For the echo in giving voice to things returns to us with vague shapes and surfaces. The ancient theory of vision which conceived of a ray proceeding from the eye to the object and back again is more literally true for the sounding echo's ability to give voice to shapes and surfaces. The blind man, who has learned and listened more acutely than we, produces this auditory "ray" with his clicking cane. Yet anyone who listens well may hear the same.[13]

I repeat the experience of the blind man, carrying with me a clicking device. As I move from the bedroom to the hall a dramatic difference in sounding occurs, and soon, as I navigate blindfolded, I learn to hear the narrowing of the stairs and the approaching closeness of the wall. Like the blind man I learn to perceive auditorily the gross presences of things. But in the relative poverty of human auditory spatiality I miss the presence of the less gross things. I cannot hear the echo which returns from the open-backed Windsor chair, but I do discern the solid wall as a vague presence. Yet in a distance not too far from human experience, I know that the porpoise can auditorily detect the difference of size between two balls through his directed echo abilities, a difference which often escapes even the casual glance of a human.

I listen more intently still. The echo gives me an extremely vague surface presence. I strike it and its surface resounds more fully. Yet even in the weakness of the echo I begin to hear the surface aspects of things. I walk between the Earth Sciences building with its concrete walls along the narrow pathway bounded on the other side by the tall plywood walls fencing off the construction of the new Physics building. In the winter the frozen ground echos the click of my heels, and I soon know when I have entered the narrowness of the pathway. Once at the other end the sound "opens up" into the more distant echoing of the frozen ground which stretches to the parking lot. But as the days go by and I listen, I soon learn that not only is there a surface presence, not only is there the "opening" and the "narrowing," but there is also a distinctly different echo from the concrete wall and the plywood fence. The surface-aspect only gradually becomes less vague in the sharpening of our listening abilities. In the echo and in the striking of the thing, I hear surfaces as existential possibilities of listening.

While there is no question here of exhausting even the relative and often vague "poverty" of shape and surface aspects, the march towards the "richness" within sound must continue. It is with a third spatial signification that this "richness" begins to appear, for, stronger than shapes and more distinct than surfaces, I *hear interiors.* Moreover, it is with the hearing of interiors that the possibilities of listening begin to open the way to those aspects which lie at the horizons of all visualist thinking, because with the hearing of interiors the auditory capacity of making present the *invisible* begins to stand out dramatically. To vision in its ordinary contexts and particularly within the confines of the vicinity of mute and opaque objects, things present themselves with their interiors *hidden.* To see the interior I may have to break up the thing, do violence to it. Yet even these ordinary things often reveal something of their interior being through sound.

A series of painted balls is placed before me. Their lacquer shines, but it conceals the nature of their interiors. I tap first this one, and its dull and unresounding noise reveals it to be of lead or some similar heavy and soft metal. I strike that one, and there is no mistaking the sound of its wooden interior. The third resounds almost like a bell, for its interior is steel or brass. In each case the auditory texture is more than a surface presentation it is also a threshold to the interior.

I am asked to hang a picture in the living room. Knowing that its weight requires a solid backing, I thump the wall until the hollowness sounding behind the lathed plaster gives way to the thud which marks the location of the stringer into which I may drive my nail. What remained hidden from my eyes is revealed to my ears. The melon reveals its ripeness; the ice its thinness; the cup its half-full contents; the water reservoir, though enclosed, reveals exactly the level of the water inside in the sounding of interiors. Hearing interiors is part of the ordinary signification of sound presence and is ordinarily employed when one wishes to penetrate the invisible. But one may not pay specific attention to this signification as the *hearing* of interiors unless one turns to a listening "to the things themselves."

In the movement from shape-aspects to surfaces to interiors there is a continuum of significations in which the "weakest" existential possibilities of auditory spatial significations emerge.

In all of this listening there is a learning. But that learning is like that of the blind man first being given sight; he does not at first know what he sees. Neither do we know what we hear, although in this case what is to be heard lies within the very familiarity with things in their present but often undiscovered richness. But once we learn to hear spatial significations, the endless ways in which we hear interiors comes to mind. We hear hollows and solids as the interior spatiality of things. We hear the *penetration* of sound into the very depths of things, and we hear again the wisdom of Heraclitus, "The hidden harmony is better than the obvious."[14]

In the reverberation of a voice given to things by the striking of one thing by another, in the echo which gives a voice to things, and in the penetration which exceeds the limits of visible space is experienced what is possible for listening. Its presence may occur in the turning of an ear. I go to a concert, and the orchestra plays before me. Suddenly the auditorium is *filled* with music. Here, Baudelaire noted that music gives the idea of space. For now the open space is suddenly and fully present, and the richness of the sound overwhelms our ordinary concerns with things and directions. But even here there lurks just behind us the relative emptiness and openness which the echo reveals. I turn my head sidewise as the music pours forth, and suddenly, dramatically, I hear the echo which lay hidden so long as the orchestra enveloped me with what is sounding before me. And in the echo I hear the interior shape of the auditorium complete even to its upward slant to the rear. The echo opens even filled space, and in hearing there is spatial signification. But let each person listen for himself.

1. Søren Kierkegaard, *Either/Or*, trans. David F. Swenson, vols. I and II (Garden City: Doubleday, 1959),Vol. I, p. 55.
2. Ibid., p. 66.
3. Ibid., p. 67
4. Edmund Husserl, *The Phenomenology of Internal Time Consciousness*, trans. James Churchill (Bloomington, Ind.: Indiana University Press, 1964), p. 149 (italics mine).
5. Erwin W. Straus, *Phenomenological Psychology*, trans. Erling Eng (London: Tavistock Publications, 1966), p. 4.
6. P. F. Strawson, *Individuals* (London: Methuen, 1971), p. 65 (italics mine).
7. Straus, *Phenomenological Psychology*, p. 7.
8. Maurice Merleau-Ponty, *The Visible and the Invisible*, trans. Alphonso Lingis (Evanston, Ill.: Northwestern University Press, 1968), p. 212.
9. J. M. Heaton, *The Eye: Phenomenology and Psychology of Function and Disorder* (London: Tavistock Publications, 1968), p. 42.
10. Ibid., p. 42.
11. Ibid., p. 43.
12. Pamela K. Haughawout, " 'I See' Said the Blind Man" (unpublished paper, 1969), p. 4.
13. Echo-location by clicks is more accurate than by tones. "If continuous tones were used instead of clicks there was a significant loss of accuracy in the perception of the direction of the sounds, though the experiences had the same character of wave lengths greater than 2K." Georg von Békésy, *Experiments in Hearing*, trans. E. G. Weaver (New York: McGraw-Hill, 1960), p. 287.
14. Philip Wheelwright, *The Presocratics* (New York: Odyssey Press, 1966), p. 79.

CHAPTER SIX

THE AUDITORY FIELD

We listen first to things. They capture our attention in their voices and are the "naïve" or first existential sources of the sounding which we hear. Yet without forgetting this first presence of the existentiality of the thing, the concern of phenomenology must also be expanded beyond any exclusive concern with things alone. To simply take the thing alone without raising the wider question of how things present themselves in terms of a situated context is to allow the illusion of a thing-in-itself to occur. The thing never occurs simply alone but within a field, a limited and bounded context.

The question of an *auditory field* has already been proximately anticipated in the observation that all things or occurrences are presented in a situated context, "surrounded" by other things and an expanse of phenomena within which the focused-upon things or occurrences are noted. But to take note of a field as a situating phenomenon calls for a deliberate shifting of ordinary intentional directions. The field is what is present, but present as implicit, as fringe which situates and "surrounds" what is explicit or focal. This field, again anticipatorily, is also an *intermediate* or eidetic phenomenon. By intermediate we note that the field is not synonymous with the thing, it exceeds the thing as a region in which the thing is located and to which the thing is always related. But the field is also limited, bounded. It is "less than" what is total, in phenomenological terms, less than the World.

The field is the specific form of "opening" I have to the World and as an "opening" it is the particular perspective which I have upon the World. It is an existential structure in that all things which present themselves do so within the field—the field "transcends" things—but the field itself is not synonymous with World. The question of World in the full ontological sense, however, arises only fully with the question of horizons which in turn "surround" and "transcend" the field of presence.

Thus the field as an intermediate eidetic phenomenon is an existential structure, no longer at the level of things as such but not yet that which allows a comprehensive apprehension of the World as lifeworld to emerge. That apprehension occurs only at the limits of phenomenology. But the second level of investigation calls for a preliminary survey of the shape and structures of the auditory field as a type of "opening" to the World.

In isolating field characteristics the temporary suspension of the first existential attention toward things must occur. Attention is turned to what is indirect and implicit when compared to the ordinary involvements with focal things. Phenomenological attention moves outwards, recapturing a possibility of the focus-fringe ratio anticipated in the first approximations to the field phenomenon. But this move away from things in order to explicate and describe the field phenomenon does not abandon the existential possibilities of things so much as it performs its purposeful inversion of attention in order to return to a more adequate sense of existentiality once the field is described.

Beginning once more with a device, an approximation as first introduced in a "visualist" form of certain features of intentionality, the question now becomes one of the auditory dimension. Once more the abstract device of pairing approximate fields offers an initial entry into what must become a more comprehensive survey of auditory field characteristics.

When the question of paired field phenomena is raised, there appear a number of *functional* similarities concerning the relative forms of the visual and auditory "openings" to the World. I note comparatively that those experiences which reveal the structure of focus to fringe with a variable ratio of relativity between them occur auditorily as well as visually. In listening to a symphony, if for some special reason I care to do so, I find I can focus upon the strains of the oboe in spite of the louder blaring of the trombones (at least within limits). The city dweller hears the clink of the coin on the subway platform even as the train approaches, and the jungle dweller hears the whisper of the adder in spite of the chatter of monkeys and parrots. I can select a focal phenomenon such that other phenomena become background or fringe phenomena without their disappearing.

Moreover, this attention is keyed into the personal-social structures of daily life in such a way that there are habitual and constant patterns of appearance to those things which normally remain fringe phenomena and those which may be focal. I go to the auditorium, and, without apparent effort, I hear the speaker while I barely notice the scuffling of feet, the coughing, the scraping noises. My tape recorder, not having the same intentionality as I, records all these auditory stimuli without distinction, and so when I return to it to hear the speech re-presented I find I cannot even hear the words due to the presence of what for me had been fringe phenomena. The tape recorder's "sense data" intentionality has changed the phenomenon.

In daily affairs this focus-fringe ratio constantly shifts with interests and occasions. In this the variability of focal intentionality functions "like" its visual occurrence. But when the *shape* of that focus is noted, there immediately appear certain striking differences as well. Within the visual field, focus displays itself as a central vision within the field. To turn my focus, I turn my eyes, my head, or my whole body. The visual field, moreover, displays itself with a definite *forward oriented* directionality. It lies constantly before me, in front of me, and there it is fixed. As a field relative to my body it is *immobile* in relation to the position of my eyes which "open" toward the World. Noted also was the vague, though noticeable horizon which imparts a *roundish shape* to the visual field. Thus as a field, the visual "opening" to the World has a concretely directed and determined spatiality relative to bodily position.

When this set of determinations of the visual field is paired with that of the auditory field, the differences of dimension begin to occur. First, the auditory field as a *shape* does not appear so restricted to a forward orientation. As a *field-shape* I may hear all around me, or, as a field-shape, sound *surrounds* me in my embodied positionality. I am sitting at my desk, and I hear my wife approaching up the steps. She enters the study and speaks to me from the doorway to the left and behind me. I turn to greet her, but she has first been present and noted from behind in the sounding of her feet. I catalogue my auditory experiences and note that the ticking of the clock comes from the right, the hiss of the radiator from the left, the hum of the light from above, and the wag of Josephine's tail from under the table. All of these sounds occur simultaneously and "fill" the auditory field with their complex multiplicities.

I also note that I can switch my focal, auditory "ray" from one sound to another without even turning my head. I can discern that the sound from the right and behind is that irksome whine of the stove fan motor.

My auditory field and my auditory focusing is not isomorphic with visual field and focus, it is *omnidirectional*. In the shape of the auditory field, as a surrounding thing, the field-shape "exceeds" that of the field-shape of sight. Were it to be modeled spatially, the auditory field would have to be conceived of as a "sphere" within which I am positioned, but whose "extent" remains indefinite as it reaches outward toward a horizon. But in any case as a field, the auditory field-shape is that of a surrounding shape. This shape may often be quite dramatically located and discerned when the field is most *full*, as in the presence of a full and dramatic moment of symphonic music. If I hear Beethoven's Ninth Symphony in an acoustically excellent auditorium, I suddenly find myself *immersed* in sound which *surrounds* me. The music is even so *penetrating* that my whole body reverberates, and I may find myself absorbed to such a degree that the usual distinction between the senses of inner and outer is virtually obliterated. The auditory field surrounds the listener, and surroundability is an essential feature of the field-shape of sound.

But if the dramatic presence of symphonic music reveals at a stroke what the poorer cataloging of separate direction possibilities also shows in regard to the omnidirectionality of sound presence, this dramatic presence can also hide another aspect of the auditory field. For while one essential possibility of the auditory field is the filling of its spatiality as in the case of dramatic music, there are other times when there occurs a *relative* emptiness.

I go for a walk and stand in the middle of a vast park in the north of London. To one side there is a roadway now filled with the evening traffic. The honks and roar of lorries, buses, and small but exceedingly noisy European cars fill *that side* of the auditory field. The other side is *almost* silent, or at least relatively quiet. For while the field of sound surrounds me, it does not do so with anything like a constant homogeneity. Visually, however, if I consider color as a constant variable, I find no "place" in the entire visual field which is empty or even relatively empty. The colorful plenum of the visual field remains constantly full. But the relative and contrasting quiet to "one side" of my auditory field presents a shifting of sound within that field such that what is "full" and what is "rarified" may variably flow within the omnidirectionality of the field as an overall possibility.

This nonhomogeneity, however, is most precisely located at the other end of the spatial signification of the auditory field in the experience of *directionality*. For whether or not I am correct about the source of the sound, and in spite of the syncopation of the visual and auditory appearances of the thing when the distance is great enough, there is the clear phenomenon of directionality within the auditory field. I hear the car coming from behind me, and I jump to avoid it. The jays calling are doing so from the direction of the locust tree. Even in the presence of the orchestra the cough comes from the right.

The clear directionality of sound, however, is not always recognized in our speech. It is recognized insofar as it retains its proper naïveté while embedded in a concern for things. "The bird is over there, I heard him call." "He must be behind the house, I heard him working there." Yet until very recent times we have not accurately determined this language in terms of directional spatiality to the degree which visual language allows. But the invention of amplification and sound reproduction instruments has begun to make this language usable. Speakers of philosophical language, which is often as sedimented as ordinary language, might balk at the usage which describes one sound as being *to the right of* or *to the left of* or *above* or *below* another sound. Yet experientially sounds may be discerned in just this way, and the avid stereo and electronics fan already uses this descriptive language. We no longer find it odd that the sound from the right (speaker) is flawed or out of balance with that from the left, and the sounds from a demonstration record which march across the room are sounds which *move from the left to the right.* The field-shapes of sound include both directionality and surroundability.

Here an enigma of the auditory field emerges from these two dimensions of field spatiality; for the global, encompassing surroundability of sound, which is most dramatic and fully present in overwhelming sounds and the often quite precise and definite directionality of sound presence which is noted in our daily "location" of sounds, are both *constantly co-present.* For the description to be accurate, both surroundability *and* directionality must be noted as co-present. This "double" dimensionality of auditory field characteristics is at once the source of much ambiguity and of a specific richness which subtly pervades the auditory dimension of existence.

A closer examination of the bidimensionality of auditory field-shape shows that there is a certain variability which auditory focusing can reveal in relation to the co-presence of surroundability and directionality. The contrast of the musical experience with everyday listening points to two such variations of focal attention. Quite ordinarily, sounds are taken directionally. The hammering from next door is heard *as from* next door. The sparrow's song in the garden presents itself *from* the garden. But if I put myself in the "musical attitude" and listen to the sound as if it were music, I may suddenly find that its ordinary and strong sense of directionality, while not disappearing, recedes to such a degree that I can concentrate upon its surrounding presence.

Contrarily, when listening to the orchestra and in the highest moments of musical ecstasy, I can (perversely, perhaps) by an act of will also raise the question of directionality; and while I continue to be immersed in the sound, there also emerges a stronger sense of direction.

Both these dimensional aspects of auditory presence are constant and co-present, but the intentional focus and the situation varies the ratio of what may stand out. There is also a noematic difference in relation to what kind of sound may most clearly present itself as primarily surrounding and primarily directional without losing its counterpart. In his experiments with hearing, Georg von Békésy has shown that the sense of directionality is much more precise with clicking sounds than with tones. Constant tones, even modulating tones, show forth more dramatically the encompassing and less directional presence of sound. The clicking "language" of the porpoise, the tapping of the blind man's cane, the ping of sonar for directionality and location are not accidental but learned selections from the realm of sound for the type of sound appropriate to the highest degree of directional intentional fulfillment. Contrarily, the use of melody, tone, and the flow of music enhances the purposeful seduction of musical presence.

Both these qualities of sound are used simultaneously in what is a most normal human activity, *face-to-face speech.* The other speaks to me in the "singing" of the human voice with its consonantal clicklike sounds and its vowel tonalities. It is a singing which is both directional and encompassing, such that I may be (auditorily and attentionally) *immersed* in the other's presence. Yet the other stands *before* me. Speech in the human voice is between the dramatic surroundability of music and the precise directionality of the sounds of the things in the environment.

It is in this range of variable presence and focus that the distance between musical experience, often taken as an exceptional experience, and the experience of sounds as primarily the sounds of things which are "located" in a place appears. The seductivity of a "pure sensuosity" in Mozart's music described by Kierkegaard finds support, but with a different ground here.[1] In the overwhelming presence of music which fills space and penetrates my awareness, not only am I momentarily taken out of myself in what is often described as a loss of self-awareness which is akin to ecstatic states, but there is a distance from things. The purity of music in its ecstatic surrounding presence overwhelms my ordinary connection with things so that I do not even primarily hear the symphony as the sounds of the instruments. But the flight of music into ecstasy is quickly lost if the instrument intrudes as in the case of having to listen to the beginner whose violin squeaks and squawks instead of sounding in its own smooth tonality.

This ecstasy is also the occasion for an illusory phenomenon, the temptation toward the notion of a pure or disembodied sound. In the penetrating totality of the musical synthesis it is easy to forget the sound as the sound of the orchestra and the music floats through experience. Part of its enchantment is in obliteration of things. A counter-variable illustrates this: a philosopher friend who now knows he is going deaf told me that he first noticed this ailment when he experienced loss of interest in music. He described the music as becoming "distant . . . objectlike . . . over there apart from me." It had begun to lose its surrounding, penetrating quality for him.

There is, however, a positive point to be made as well concerning the experience of musical ecstasy and the way in which musical sound does form a gestalt. What disappears in the symphonic presentation is the sense of the separate and discrete "individuals," at least in a relative sense. The "instrument" which sounds is the entire orchestra united in sound. The surrounding, penetrating quality of sound maximizes larger unities than individuals as such.

Conversely, the hunter intent upon bagging his game misses the musical sonority of the bird song, not because it isn't there, but because it is the direction and location of his prey which motivates him. So, too, with most daily concerns, directionality is that which stands out and is sufficient for ordinary affairs. The continued attempts to enhance musical surroundability continues in technological society in the move from two- to four-channel stereo production in the hope of embodying even more fully the omnidirectional surroundability of musical sound, while in the refinements in all types of echo-location down to the sonic probes of the earth itself or the sonic probe of a diseased eye with a minisonar to discover a detached retina, there occurs the precise determination of directionality and shape.

It is in the ordinary babbling traffic which we have with others where the ambiguous richness of sound is both directional *and* encompassing that there is revealed a special kind of "shape." This is what may be called an auditory "halo" or the *auditory aura*. The other, when speaking in sonorous speech, presents himself as "more" than something fixed, "more" than a outline-body, as a "presence" who is most strongly present when standing face to face. It is here that the auditory aura is most heightened.

The experience of an auditory aura is "like" the experience of music in which intentionality though keenly aware, "lets be" the musical presence so that the sound rushes over and through one. But it is not like music in that the temptation to become disembodied, to allow oneself to float away beyond the instrumentation is absent. Rather, in the face-to-face speaking the other is there, embodied, while exceeding his outline-body, but the other is in my focus as there before me face to face. It is in his speaking that he fills the space between us and by it I am auditorily immersed and penetrated as sound "physically" *invades* my own body.[2]

Moreover, the ambiguity of the auditory aura, most vivid when I am directly facing the other, is also part of the way in which hearing within the auditory field is structured. I listen for sounds with attention to direction. I sometimes find that there is a 180-degree error: I momentarily mistake the direction of a sound coming from in front of me to be one coming from the back, or vice-versa. Ordinarily the mistake may be quickly corrected, usually by turning the head briefly and thus allowing the inadequate, momentary perception to be adumbrated more adequately. But in the moment when I stand within the ambiguity of the fore-back direction, I also discover the possibility of the face-to-face aura presence which is a subtle existential possibility of the other as an auditory presence. For not only may the field be relatively "empty" on one side or the other, but there is a discernible difference between the listening which occurs face to face and that which comes from a side. Were we again to re-enter the construction of a purely reflective way of discovering the "shape" of our auditory perceptual opening to the world, this would provide one such clue for a reflective self-recognition.

The presence of the other embodied auditorily in the "excess" of the aura which not only "exceeds" the presence of the outline-body, but "fills" the space between us is yet another instance of the experience of the *invisible*. It is in the *voice* that the "excess" is *heard*, and a full sense of the presence of things and of others is one which calls for such listening.

The auditory aura is, of course, by no means restricted to the face-to-face speaking situation. It is present throughout the range of auditory experience, though not always so notably as in the face-to-face situation. Listening to music often may reveal the auditory aura as well, but it is best located by actions which disrupt the presumptive ideality of musical listening. The well-built auditorium "conceals" those auditory features which disrupt the immersion in the music. Nor is it accidental either auditorily or visually that the audience "faces" the orchestra, or that the better seats are those which are closer to the center. For when I listen to music I also *face* the orchestra, and the richness of its aura is such that while facing the orchestra the plenum of sound is full and penetrating. But, as noted above, when I begin to engage the movements of my body which I ordinarily use to locate directions and do so extremely enough, I can suddenly discover the echo from the back of the auditorium which vividly disrupts the previously full "halo" of the music.

To this point the auditory field has been surveyed with an ear to its spatial field significations. Within this field plenum of sound the range of variability from the rich and full filling of the field to the discernment of precise direction within the field reveals something of its "Parmenidean" features. Parmenides, as an ancient philosopher of presence, characterized Being as a "whole," "without end," "one," "continuous," and "homogenous or filled within the limits." He characterized the limit thusly, "Being is complete on every side, like the mass of a well-rounded sphere."[3] A phenomenology of auditory field presence rediscovers these characteristics of experience.

But the spatial existential possibilities of sound do not exhaust its invariant features. Within the field plenum of sound there is also a "Parmenidean" *continuity*. So long as I experience, there is a perceptually continuous sound presence. For although any one sound or even all sounds may be at the very edges of my consciousness, they are not totally absent, nor can I escape sound. Throughout the day the ebb and flow of noise is continuous. If it is loud, as in the factory, the airport, or even the city, I escape to the quiet of the countryside and I notice the "silence"; but strictly speaking this is quiet, not silence. The rustling of the wind in the trees is quieter than the rush of the subway train, but it is a sounding. Even in the desert there is the wind and the crackling of the sun on the sand. And in the ultimate "escape" from noise in the anechoic chamber, I am suddenly startled by the noises of my own body which lie masked in daily affairs. My breathing, the "whine" of my nervous system, and the inhibited flow of my bloodstream suddenly appear in the quiet as noise.

Continuity of presence is not restricted to auditory presence but is a field characteristic of all perceptual experience. I continue to "see" even when my eyes are closed, for, while I have closed out the things before me, my field does not become empty or disappear but merely turns dark (black or reddish). Fields as fields are constant presences which are never empty but filled as a plenum. Sound is continuously present to experience.

But this presence is also a *penetrating*, invading presence. Sound penetrates my awareness. As noise this penetrability may be shattering, ultimately even painful. The sudden scream at the moment of highest tension in the Hitchcock movie upsets my composure, and it is rightfully described as *piercing*. The sound of the siren coming from down the street sometimes makes me cover my ears to escape its intensity, but even then I only slightly muffle it. The rock concert in its musical orgiastic decibles takes me to the very threshold of painfulness at points. And in the Orwellian fantasies which now seem to be coming true, the police and political powers consider the development and use of high-intensity sounds to quell riots.

This existential possibility of sound has long been noted. The languages which relate hearing to the invading features of sound often consider the auditory presence as a type of "command." Thus *hearing* and *obeying* are often united in root terms. The Latin *obaudire* is literally meant as a *listening* "from below." It stands as a root source of the English *obey*.[4] Sound in its commanding presence *in-vades* our experience, and although this invasion may be desirable, as in the cases of musical enchantment, it may also be detestable as in the unwanted noise of the jackhammer early in the morning. In both cases one's train of thought is likely to be upset by the "command" of the sound which is so penetrating or loud that he can't "hear" himself think. The ability to reveal interiors, as the essential penetrability of sound presence, even applies to myself as an embodied being. Sound physically penetrates my body and I literally "hear" with my body from bones to ears.[5]

If, noematically, sound penetrates, noetically there is also the problem of a response to this continuous invasion. In ordinary vision unwanted sights may be often simply closed out by shutting one's eyes. And although in the presence of intense lights, as in trying to sleep on a transatlantic flight in the presence of the movie screen, such simple measures do not suffice, a set of opaque sleep blindfolds is sufficient to close off the visual disturbance. But not only do our ears have no flaps to close off the sounds, sticking the fingers in them fails to solve the matter. Ultimately to escape unwanted noise we have to either actually remove ourselves from its vicinity or build a protective environment which shuts it out (the ordinary house is not sufficient protection to close out the sound of the jackhammer). This penetrability invades even the mute objects around us. This strength of sound is, of course, one of the factors which is also one of its weaknesses noted above. The penetration of interiors is precisely that possibility which exceeds the clear stopping at the surface which vision allows in some of its richer spatially distinct features. But the relative resistance of the thing does allow some sense of difference: there are auditory correlates to transparency, translucency, and opacity.

In terms of sound penetrability, however, the escape from or control of sound is essentially a matter of psychic control. I may even become habituated to loud sounds to the point that they do not count for the same disturbing invasions of myself which they at first show. The factory worker learns to tune out the machinery. And the youth-cultist seems even to thrive in the presence of noise.

The auditory field, continuous and full, penetrating in its presence, is also *lively*. Sounds "move" in the rhythms of auditory presence. Here we approach more closely that first listening which detects in sound an essential temporality. The fullness of auditory presence is one of an "animated" liveliness.

This existential possibility of sound ranges from the most abstractive to the most ordinary to the most extraordinary experiences involving sound, and it is often pointed out in the contrasts which occur in the absence of lively sound. Not only is Notre Dame without the choir empty, but it may suggest even a certain deathlike quality. In contrast, when sound is added to abstract figures, they "come alive." I go to a movie, and a short cartoon feature precedes the main show. It consists of black and red dots which already start to "come alive" when they begin to move across the green ground. But as they bounce off one another, jumbled noises mimicking speech are presented. Suddenly the moving dots in their lines and bumpings, in the presence of the "speech" and sounds of "yelling" become soccer players seen from above. The dots are animated by the sound which makes their motion anthropomorphically understandable.

Conversely, the sudden absence of sound can disembody a scene. In the movie *The Battle of Britain*, a technicolor reenactment of the air battles over England during World War II, at the height of the decisive battle Spitfires and Hurricanes dance in the air in combat with Messerschmitts and Junkers. Amidst the loud chatter of the machine guns and the roar and sputter of the airplanes the sound track is suddenly and deliberately silenced. At the instant of the disappearance of animating sound, the scene becomes eerie, a moving tableau which becomes more abstract and distant. This momentary irreality of the disengagement of sound allows the battle to be seen as a strange dance without music. Emptiness which can be uncanny is silence in the auditory dimension.

The cinema films, those concrete exercises of phenomenological variations, provide endless examples of this same animating quality of sound.

In *2001: A Space Odyssey*, it was found that without the sound of background music the slow drifting of the spaceship did not appear even as movement. Also, the old favorite, the silent movie, is accompanied by the piano. The intimate relation between animation, motion, and sound lies at the threshold of the inner secret of auditory experience, the *timefulness* of sound. The auditory field is not a static field.

Here, then, we reach the completion of a first survey of the field-shape of sound. It is an omnidirectional "sphere" of sound which is variably full and/or rarified in a ratio of relativity. This same ratio of relativity pertains to the co-presence of the "shapes" of surroundability and directionality, manifestations of sound presences. The field of sound is also a penetrating presence which in certain instances unites and dissolves certain presumed "individualities" by its penetration in and through interiors in a power of penetration. This power of sound is also a dynamic and animated or lively quality of sound. And while all these existential possibilities of the auditory field are present in sound, dramatic and selected variables reveal these qualities in more striking form.

1. Søren Kierkegaard, *Either/Or*, trans. David F. Swenson, vols. I and II (Garden City: Doubleday, 1959), 1:55.
2. Georg von Békésy, *Experiments in Hearing*, trans. E. G. Weaver (New York: McGraw-Hill, 1960), pp. 172–73.
3. Philip Wheelwright, *The Presocratics* (New York: Odyssey Press, 1966), p. 98.
4. Erwin W. Straus, *Phenomenological Psychology*, trans. Erling Eng (London: Travistock Publications, 1966), p. 287.
5. von Békésy, *Experiments in Hearing*, p. 164. "It was found impossible to construct an earplug whose attenuation for air borne sound was more than 35–40 db in the frequency range between 100–8,000 cps. Some authors . . . have attributed this limitation of performance to the vibrations set up directly in the bones of the head by the sound field."

CHAPTER SEVEN

TIMEFUL SOUND

The tradition concerning the experience of sound is one which situates hearing as the temporal sense and the "world" of sound as one of flux and flow. The postponement of a consideration of this temporal movement of sound in order not to bypass the spatial significations of the auditory dimension must now give way to the examination of what appears first in the reflections upon sound. Sound dances timefully within experience. Sound embodies the sense of time.

I listen to the presentations of sounds. The oliveback thrush issues his song which is punctuated by the less obvious crack of a twig in the forest. Both these sounds are accented by the interludes of quiet which are filled only with a breeze or the rustle of leaves. Each of these sounds comes into being and passes from being in a temporal dance which does not submit to my will.

At the first existential level, of course, these sounds are experienced as the sounds of things. The forest dweller does not confuse the beaver's distinctive splash with that of the trout's leap. Nor is there at this level a sheer sense of buzzing confusion or of structureless flux, unless the listener is exposed to a strange world not yet fully heard. For the flux and flow of auditory experience, upon a more concentrated listening, displays its own deeper rhythms and harmonies. The shapelessness of an initial flux and flow is due to the level of reflection, not to the existential possibilities of sound.

Here again we note that a phenomenological listening is also a learning which allows the phenomena to more and more clearly show themselves. The deeper significations, not absent from first listening, do not "show" themselves fully in first listening. The level of reflective experience can easily be confused with what the phenomena show. For example, first investigations into perception may be noted to undergo definite stages of development. The observations of flux and flow during a first reflection upon experience is not restricted to listening, although it is more dramatic in relation to each of the "senses" than it is for sight. But this is to be expected if sight is both a culturally preferred sense, and if its possibilities of experience are more thoroughly sedimented in a tradition of interpretation. It is harder for us to immediately relearn or regrasp the movement from ambiguity to polymorphous but structured wealth. It is hard for us to return to the "first seeing" of the blind man who also first sees colors which seem to fill the space of the room and "jump out at him" in a visual "flux and flow."[1] Students, for instance, doing a beginning phenomenology of some perceptual dimension almost always are struck by the occurrences in the process. First, they note that there is often a lack of clearly experienced subject-object distinctions. In tactile experience to lie on one's backside and to have to describe the experience calls for a noting of the indefiniteness and vagueness of where I end and the couch begins. But to leave the investigation here is to confuse an initial apodictic level with adequacy: the expert safecracker knows quite well how precise touch can be in his experience of tactile objects. Secondly, the beginner notes a certain struggle with language (which is also a struggle with experience), for he comes upon phenomena which he has not previously noted and does not yet have or call to mind the words with which to describe such experiences. He is called upon to provide a description of nuances which produces first metaphors or constructed terms before he can find the appropriate terminology. Thirdly, as the reflective process continues, the beginning investigator begins to find that he gains a progressively finer sense of discrimination concerning the things which he experiences.

Such a progression is not unknown in ordinary experience even if not thematized as a reflective phenomenology. The beginning wine taster, after all, can tell little about the differences between the ninety-eight-cent fortified wine and the 1964 Medoc; in fact, for him, all wines taste alike, a statement which horrifies the accomplished wine taster who has forgotten his own early experience. The inability to discern the difference between a delicate sauce and plain gravy is characteristic of the quick-snack artist, not of the gourmet. Thus the question of temporality of sound cannot be left with the correct but superficial sense of the intimacy of temporality within sound.

To return to listening with an ear to the time and times of sound also calls for variations upon more ordinary listening patterns. The first existential level which attends to the sounds of things, particularly as the sounds of things primarily in the central focus of ordinary interest, needs adumbration into the larger gestalts of auditory temporality. There then appears a timeful march of daily sounds which may be spoken of initially in terms of the *rhythmic* temporal movements of sound.

This daily rhythmic quality to auditory presentation is an ordering of the first flux and flow which takes place within experience in terms of the background-foreground sounds in rhythm which are part of daily life. If I am in my summer home there is no mistaking the coming of the day in the chorus of birds. The night sounds of the porcupine and the owl are different but have their own regularity. In the city there is a rhythm, too, with the rush hours and daily rising crescendo of noises which at night recedes to pianissimo only toward the morning hours.

This music of daily sound may be described analogously as having the foreground and background textures of a melody with some form of bass accompaniment. In the woods the accompaniment is the constant babble of the brook and the sighing of the wind in the trees. In technological society that accompaniment is the almost constant, pervasive presence of a form of machine-produced hum. Within the self-enclosed buildings of contemporary architecture there is the whir of the heating and air-conditioning machinery, the hum of the lighting, and the electronic whine of the technosphere which is the counterpart in that environment to the tides and the winds of the wild world.

These rhythm-section sounds are not effervescent or abrupt comings and goings of sound but an auditory texture and background which provides an auditory stability to the world. The lack of such a stable background, or a dramatic change in its rhythm, is the occasion for human anxiety. On the seashore the rhythmic splash of waves with the tide is both comforting and hypnotic, but the disruption of the hurricane or the sudden storm augers disquiet for the shore dweller. In Cambridge, Massachusetts, a number of years ago, a church installed a very advanced air conditioner. Yet the congregation continued to feel hot even though the temperature and humidity gauges indicated all was well. It was only after the engineer discovered that they couldn't hear the reassuring presence of the machinery that the problem was solved. An artificially produced fan noise soon made all feel comfortable, and the air-conditioner was "felt" to be effective.

The temporal rhythms of daily sound are *structured* rhythms, and it is in rhythm that the background or field of auditory temporality is located. As a field background, rhythm is a *repetition* which is the index for auditory "sameness" or stability. The repetitions of the "same" morning cock crow, which gives way to the "same" sounds of the town clock or noon whistle, fulfill the expectable temporal background pattern which allows humankind to take a certain stability for granted in ordinary circumstances. In this, rhythmic repetition plays a role functionally isomorphic with the stability of the ordinary visual background of immobility.

In ordinary visual experience the stability of the background against which the movements of objects may be seen remains the unthematized temporal resting or enduring of the visual world, but its temporality does not stand out. In ordinary auditory experience the equally taken-for-granted rhythms, familiar in their contextual expectability, provide the same stability function. Counter instances where the background is disturbed further adumbrate the role of this function. When ordinary visual experience is disrupted by what are usually background features undergoing dramatic changes, a certain disorientation of ordinary experience may occur. For example, in a situation in which everything within the visual experience is set in motion or obscured from its usual ratio of foreground to background, the sense of "visualist" stable spatiality breaks down.

The first time a person goes to sea, if the seas are rough and the boat rolls, the whole ship can be seen in motion *even from the inside*. Or if one is inside an airplane which begins to bank, even with the windows closed one can see the plane turn from the inside. (Physically speaking this is, of course, not merely a visual experience. The inner ear as an organ of balance is implicated in this seeing.) If the day is light enough, the line of the horizon is often sufficient to reestablish some sense of reference for stability. But when even this is obliterated the usual sense of spatial relations begins to be modified.

I take my canoe into the harbor in a heavy fog and find there is no horizon. The fog blends indistinguishably with the water which surrounds my canoe. It is almost as if I were floating in a basin whose edges seem to curve upward. If I approach a buoy, it appears through the fog as if elevated higher in the air than it should be and returns to its expected position only when I approach closely enough for it, too, to become surrounded by water.

Auditorily the same disruptive change of background familiarity into foreground familiarity or an unexpected change of its rhythms produces a similar experience. So long as the constant whir of the fan or the electronic hum is relatively low in intensity, it remains barely noticed; but the continued presence of a single, loud, and intense sound quickly becomes disorienting. It even appears to waver and modulate in my hearing, and such a continuous sound is the auditory equivalent of the constant drip of Chinese water torture.

There is also a rhythm which lies hidden in the very stability of the mute object in vision which retains its motionlessness in a ratio to the motions which I make with my eyes. If I stare at the rock, not allowing my eyes the rhythmic motion which glides over the surface of the object, the rock begins to "jump" or "waver" and ceases its normal stability; only here the ratio is inverted, for it is in the moving "grasp" of the gaze that stability is established.

There is in both visual and auditory experience a dialectic of motion to stability which undergoes various forms of relatedness. If sound is always in "motion" this motion itself in the regularities of rhythm functions as a stability. If the stable visual object is fixed by the unattended-to motion of my own eyes, this too is a ratio of movement to stability. But in each case the stability function, in ordinary affairs, takes place as the implicit and presupposed background against which the foreground themes of life take shape. Temporality is neither simple flux nor enduring stability alone.

In the discussion of the rhythms of daily life, particularly in regard to the intimacy of temporality in sound, there is a preceding analysis which has already paved the way for further variations upon sound in time. Husserl's development of a phenomenology of inner time consciousness has already opened the way for some investigation into the auditory dimension. Here, however, that analysis will be remodeled in terms of the language of focus, field, horizon, and so forth.

A phenomenology of experienced temporality soon comes upon the notion of a *temporal span* or duration of sounding which is experienced in listening. I do not hear one instant followed by another; I hear an enduring gestalt within which the modulations of the melody, the speech, the noises present themselves. The instant as an atom of time is an abstraction which is related to the illusion of a thing in itself. In terms of a perceptual field we have noted that a thing always occurs as situated within a larger unity of a field; so temporally the use of *instant* here is perceived to occur only within the larger duration of a temporal span, a living present.

Moreover, according to Husserl's prior analysis (to be modified below), this temporal span displays itself as structured according to the onset of features coming into perception, *protension*, and the phasing and passing off of features fading out of presence, *retention*. Within the temporal span the continuing experience of a gestalt is experienced as a succession within the span of duration. Thus the passing automobile whose auditory "Doppler effect" of changing pitch presents itself as a unity within the temporal span. Or when I listen to someone speak, I do not ordinarily hear a syllable at a time, or even a word, but I hear the larger melody and flow of speech as an ongoing rhythmic unity.

But by an act of deliberate concentration I find that I can so concentrate to get a syllable or a word.[2] It is, however, important to examine this attentional aspect of auditory intentionality. When I do so concentrate, seeking out a certain syllable for example, what happens in the experience is that the ordinary flow of speech becomes background and may not be grasped significantly at all. Or if it is grasped, it may have to be reconstructed in experience, as when one listens only vaguely to what someone else is saying. In such experiences lies the phenomenon of attentional intentionality which may be termed here a *temporal focus*.

In approximate terms, temporal focus operates similarly to all the focal phenomena previously noted. First, it is a focus *within* a larger unity, a field. In this case the field is the temporal span of lived-through time. As such the focus is attentional and selective in its operation so that the ratio of a foreground to background effect obtains. In the examples noted above a "narrow" instance of auditory focal attention is one which attends to single syllables such that the ordinary flow of speech becomes background. Here effects similar to those noted earlier may also be noted: the "narrower" the focus, the more the background recedes into a fringe appearance. Thus in a highly concentrated "narrow" focus I get certain sounds in the other's speech but may find it almost impossible to note what was said; and contrarily in a "broader" focus, as in attending to what is being said, I may miss or barely be aware of the aspirated *s* which is characteristic of the other's speaking style.

Here, too, the relation of *instant* to duration may be located within the experienced temporal span. In a narrow focus I auditorily grasp an instant (as foreground) within the ongoing span (as background). And in this sense there remains the variability of focal acts previously noted. The attentional focus may be narrow, fine, or broad. If now the possibilities which occur in a broadened focus are examined, there appears yet another similarity with the first approximation within focal—field presences. A maximally broadened focus is "panoramic," as for example in the case of listening to a piece of music in a relaxed mood (not listening critically to seek particular instruments or notes or themes).

Within auditory experience all these possibilities of auditory focus are temporally located as temporal focal acts also. But to this point only the *attentional* aspect of the temporal focus has been noted. There is also a question of the *shape* or temporal directionality of the temporal focus. Here it is important to note that a phenomenological investigation of temporal structures is such that a "linear" metaphor concerning the auditory-temporal field would be quite misleading. "Linearity" is a reduction of the complexity of temporal duration and *depth.* Within auditory temporality the temporal span shows itself as containing a multiplicity of auditory events which are intentionally graded. There is both a simultaneity and a succession: Eric arrives with the clink of the milk bottles which mix with the sounds of Lisa drying dishes in the kitchen. All the while the cement mixer continues its put-put sound even though it is Saturday. Within this plethora of sounds the attentional selectivity of auditory intentionality continues. I auditorily scan the multiplicity varying my attention in terms of the sounds which catch my attention or in terms of those I seek out. I may narrow or broaden this attentional aspect of auditory focus at will.

But this general feature of auditory intentionality does not yet reveal the *shape* of a temporal focus as such. If now I raise the approximate question in relation to one shape of focal perception, the question of this feature becomes clearer. In visual perception the shape of focal attention found its locus in a central core within the visual field. Certainly this focal core is variable (within limits) and in field states may be even panoramically expanded to approximate the "attending" to a field as a whole (again within limits). However, although this broadened shape occurs it must be noted that a ratio of explicit and distinct to inclusive but less distinct remains. In the field states of boredom or enthrallment while viewing panoramically I do not have a *detailed* attention, for as soon as I begin to seek out details centered focal attention may again be reflectively noted. Thus the gravitational shape of a visual focus is weighted in the center of the visual field as a phenomenological structure.

If now, "analogously," the seemingly odd question of a shape is raised in terms of temporal focusing, the question is whether such a focus is similarly bound to the "center" of a temporal span, to a now-point, or whether it displays a different set of possibilities. Is temporal focus centered toward the middle of the temporal span or does it, particularly in a closer "analogy" to the spatial shape of auditory focusing, display the possibility of movement throughout the temporal span?

Such questions at first seem unusual or even odd, yet their point soon becomes apparent in the experiences of listening which concentrate upon temporal features of the auditory field. Husserl has already characterized the temporal field as one which presents itself in terms of what are here field-like characteristics bounded by horizons. The "now" is futurally funneled by a set of expectative *protentions*; these expectations belong to our temporal experience. Protentions are the temporal "empty intentions" which "search" the coming into presence such that they may be fulfilled or frustrated. Protentions are the attentional structurings which may be futurally oriented.[3] However, it is also the case that Husserl did not deal primarily with futurally oriented protentions, and in fact his analysis of time consciousness was almost exclusively oriented toward the movement into the sense of the past. The field of time, however, is shown to contain a great deal of complexity. The coming-into-being of a perceptually temporal experience is spoken of as a 'welling up' with a 'leading edge' which Husserl often characterized as a *source-point*.[4] The other extreme of the field is a "running off" of phenomena in *retentions* which are sometimes characterized as *reverberations* or *echoes* which 'sink' into the just-past. At their extreme point there is a horizon which transforms primary retention into genuine recollection which is the first genuine appearance of memory. Husserl's own primary concern with the phenomena of "running off" and sinking back into the past in contrast to the equal concern with a futural concern may be evidenced in a modification of the diagrams which he uses for time consciousness. (fig. 5).[5]

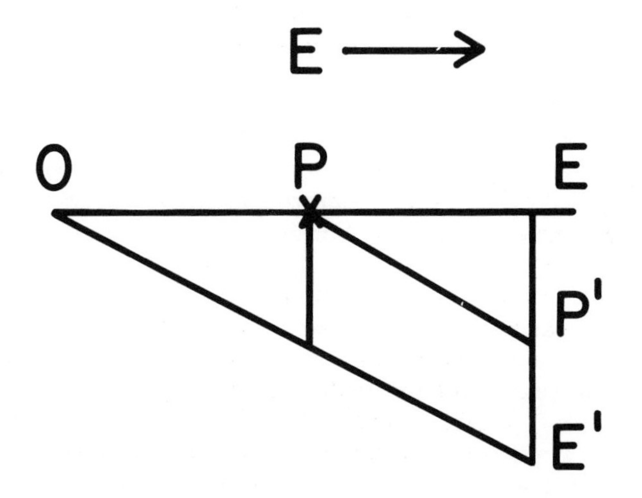

I have combined two Husserlian diagrams in figure 5. For Husserl, OE is the series of now-points, the temporal span in this context; OE' is the "depth" of the span which contains within it the reverberations and sinking back phenomena of those points which "trail off" in retention until they "disappear." They reach the *horizon* of the past, and when this occurs they may be returned under a *recollection* which is noematically and noetically differentiated from the "trailing off" of what occurs within the present temporal span. For Husserl the "direction" of time is from the future toward the past as represented in the directional arrow above the diagram. Husserl is, of course, quite aware that there is also a horizon of the future from which the original onset of what appears within time consciousness presents itself as a source-point. It is further the case that Husserl himself opens the way toward the modification which is introduced below, in that *within the temporal span* it is possible to narrowly focus upon *either* the source-point or the onset of an event (here a sound, as is often used by Husserl) or upon the phenomenon of "running off" or "trailing off." It is this capacity of temporal focus which points to certain important and characteristic aspects of the shape of temporal focal-field-horizon structures.

Prior to the modification, two simple variations which experientially establish this variability of temporal focus need be noted. If I am to be the subject of a psychological experiment in which a click is to be the signal of some action, I listen intently for that short and barely enduring sound. My protending expectation "searches" the futural "edge" of the temporal span in order to be prepared for the onset of the click. I have "pushed" all other auditory-temporal factors into the background and listen only for the click. As soon as the click presents itself at the futural "edge" of the temporal span, I no longer attend to it but as quickly as possible pass on to the act it signals. In this I specifically do not attend to its "running off" reverberation.

Conversely, I am now to listen to a tone to identify its position in the musical scale. Again I listen intently with the same selectivity for the tone. This time at its presence I do not attend specifically to its instantaneous source-point but pay special attention to its tonal quality which appears even more strongly in its "running off" reverberation, and I identify it as middle C. In these two variations I "aim" my focal attention to different aspects of the event in a variation of focus. This capacity is of extreme importance to our acquaintance with auditory phenomena.

The first modification to a merely ordinary focus, then, is one which must take account of focal shifts within the enduring temporal span. Were Husserl's diagram to be used, two essential possibilities of temporal focus could be located thusly: (fig. 6)

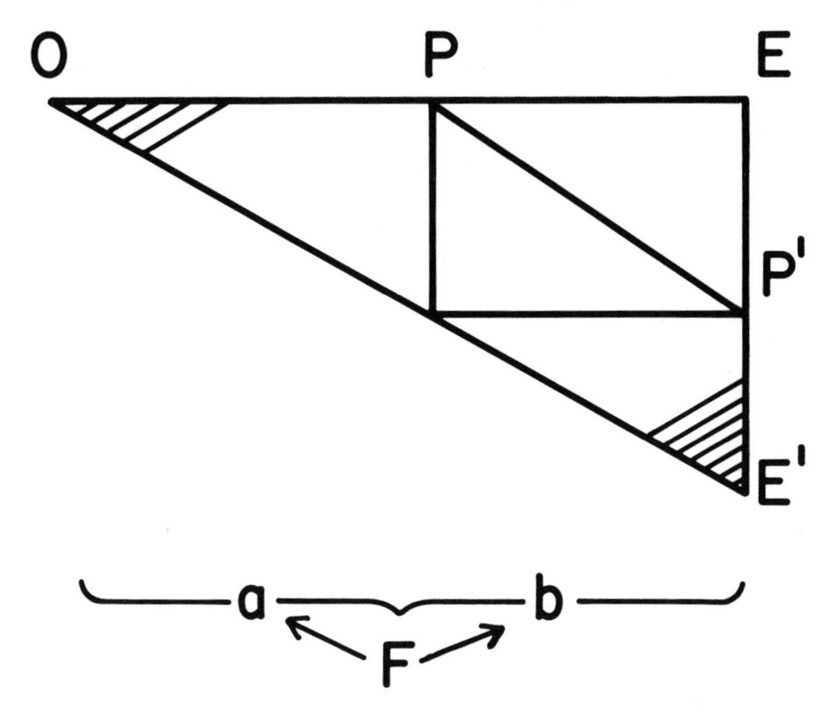

Where F is the (noetic) act of focus its aim may be either at the "leading edge" of the temporal span directed toward the onset of the source-point of the auditory event (a), or directed toward its reverberation as the "trailing off" of the sound presented (b). Here once again we have a relative variability to the focal act. Furthermore, the variability displays the same ratio of focal-to-fringe inversions as all other acts of focus. If I attend to the reverberation or "running off" of the sound, this in no way obliterates the coming into presence of the continua of auditory events, but it does allow them to become *relatively* fringe-like. Inversely, when I attend to the onset of each source-point, say a new note in the musical phrase, the reverberation in "running off" does not disappear but is relatively fringe in ratio to the focal attention. Finally, it should be pointed out that there is a sense in which this variability in each of these variations is already a highly concentrated and "narrowed" focus in contrast to the more ordinary listening in which one is usually engaged. But in this case the "place" of the narrowness within the temporal span is located either at the oncoming source-point or at the trailing-off reverberation. A third possibility is, of course, a broad focus which extends throughout the temporal span. In listening to music, for example, one usually allows the full richness of the musical presence to occur in what is here a *broad* or *open* focus with the onset of each note enriched by the depth of those which have just preceded it "equally" present.

There is, moreover, a certain peculiar "privilege" for certain types of phenomena in the location of such features of temporality. If I return to the variations upon the mute, stationary object in contrast to various types of "moving" phenomena, I now more precisely locate one of the sources of the characterization of time as a type of "motion." If I look at the calendar on the wall, it stands out as motionless and mute, and in relation to it I detect only a massive *nowness.* Its appearance neither dramatically *comes into presence* nor *passes from it* in its motionless state. If I want to take note of its "temporality" I must already make a reflective turn to *noetic* phenomena: it is *my consciousness* which is *aware* of the *passing of time* before this object. However, if the object is moving—my son's baseball suddenly looms before me, and I must either catch it or avoid it before being hit—in the duration of the event of the ball coming toward me the moving ball allows a shift toward the *noematic* appearance of successive time. But even more dramatically, when I listen to auditory events there seems to be no way in which I can escape the sense of a "coming into being" and a "passing from being" in the modulated motions of sound. Here temporality is not a matter of "subjectivity" but a matter of the way the phenomenon presents itself. I cannot "fix" the note nor make it "come to stand" before me, and there is an *objectivelike* recalcitrance to its "motion." Conversely, when those rare occasions arise on which one is purposefully placed in the presence of a single, sounding tone which does not vary and in which the depth of foreground to background features is eliminated, this presence can not only be deeply disturbing, but it begins to approximate the solid "nowness" of the stable visual object and time sensing "returns" to its location "in oneself." The intimacy of sound and time appears as an existential possibility of sound which reveals its range of forms.

Focus, however, is to be located *within* a field which in turn is bounded by horizon if the initial model may be discerned in relation to temporal as well as spatial features. The question of a field poses immense problems in relation to the experience of temporality, and there is a sense in which within the limits of this study important dimensions of the temporal field must lie outside of consideration. But one modification upon the Husserlian notion of time consciousness is called for here. It is one which more equally notes the protensive futural time intention.

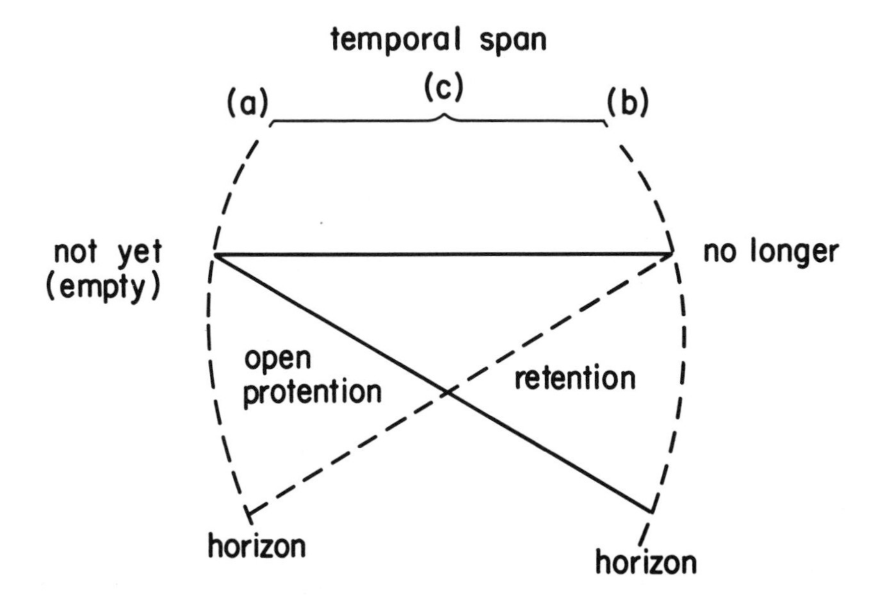

In figure 7 the span from (a) to (b) is the durational span of present time awareness. It "trails off" in retentions until the horizon is reached, at whose "edge" retention changes into recollection which can also be presented but in a different noematic modality. The *field duration* is the totality of what is or may be "within" temporal awareness (c). The horizons of the "future" as the area of the not-yet relate to protended expectations which display themselves in various types of imaginative projections (plans, daydreams, specific expectations, predictions) which, while located within a present awareness, emptily intend "toward an absent horizon." Likewise, that which is no longer presentable in its perceptual retentive fulfillment, such as those types of recollections and other constituted past intentions of "history," points toward the "other" limit of time consciousness, the horizon of the past. The interest here, in which the experience of auditory phenomena with regard to its temporal presencing, concerns only the structure of the span of temporal awareness. And although this has been designated the span of field duration, it can be seen from the previous variations that there is a sense in which within field duration the focal act can be concentrated either at the limits of the protending horizon of field duration or at the limits of the retentional horizon, or it may "span" the entire durational presence. Within the limits of field presence, focal concentration is not limited to a "center."

In passing it may be worth noting that the modifications proposed retain the "borrowed clarity" of the visual-spatial language which remains so much a part of the philosophical tradition and equally a part of a Husserlian-styled set of approximations. It might be thought that a shift toward a more musical set of terms would be helpful, but an examination of these terms often reveals precisely the same type of cross-sorting in terms of visual-spatial metaphors. One speaks of one note being *higher* or *lower* than another; music may be *bright* or *light*; there is also a *distance* between notes, and so forth. There is an *intensity* of sound, but intensity is also a term regarding light. This borrowing of extant approximations, however, has its advantage in its silent use of clear conceptuality which is sedimented in visualism.

However, now that both the attentional and directionally shaped aspects of temporal focus have been preliminarily noted, it is possible to return to the functions of this variability of temporal focus within auditory experience. It is now possible to retrace the range of spatial features and indicate the role which temporal focal (noetic) acts play in the constitution of spatial significances. In performing this descriptive analysis it appears that in regard to temporal factors a shift has occurred in the order of procedure. This is a shift from noting noematic to noting noetic correlates. But this shift is partly deceptive in that it allows tentatively some of the traditional prejudices concerning time to remain extant. If time appears as "more subjective" in temporality, that is due partially to the traditions which continue to value stability over nonstability. Here, however, stability and changeability should be essentially interrelated as they are within the limits of a Husserlian-styled first phenomenology.

Temporal intentionality is deeply implicated in every spatial signification of the auditory dimension. In a most preliminary and general way this may be noted in the manner in which a certain temporal duration gradually renders these significations clearer from the "weakest" to the "stronger" spatial significations. In the presentation of edge-shapes, for example, the auditory presentation is clearly dependent for its degree of clarity of shape upon more than an instantaneous presentation. The marble which rolls in the box presents its edge-shape only if its roll endures long enough. If the marble were to be dropped on the floor and then caught on its first bounce, the auditory result would not be adequate for even the slightest discernment of an edge-shape other than that of a point of contact. And, in this "weakest" aspect of auditory spatial awareness, the longer the duration (within limits) the more "precise" the edge-shape phenomenon.

The same is the case with the detection of surface characteristics. But here the question of how the thing is given its voice is also important. In the instance of a *contact voice* which results when one object strikes another, it has already been noted that the surface "area" which is revealed is very circumscribed. Repetition and duration are also implicated here. The footsteps in the hallway as they *repeat* themselves in a pattern of approaching nearer while also sounding on the tile through the *duration* of the repetition presents at least a series of circumscribed surface aspects. The auditory surface, in this example, does not "spread itself out" in a wholistic gestalt of entire surface but presents itself as a series of surface-aspects. This can, of course, be modified by, say, sliding an umbrella handle across the tiles so that it clicks in each crack between them. But again the surface in this case approximates an edge-shape and does not spread itself out. In all these variations the repetition and duration of the temporal dimension presents itself as "making possible" the "weak" spatial discrimination which sounds present.

A somewhat different aspect of surface is presented if the giving of voice is that of the *echo voice*. Here the distance of the surface reflecting the echo voice may be discerned within limits, and perhaps even a quite vague recognition of an internal shape is revealed as in the case of echoes in an auditorium. But the surface is presented vaguely as present and perhaps as relatively hard or less hard (concrete versus plywood walls), at least within the limits of human hearing ranges as contrasted with the more precise hearing ranges of animals. Again repetitions and durations are deeply implicated in the clarifying of such surface presences. The repeated clicks of the blind man's cane or of a constructed echo-directional sounder almost seem to temporally build up the gradually clearer sense of surface presence. With sound the subtle time dimension "allows" the gradually clearing spatial significances to be known.

Each of the above phenomena have been located phenomenologically. Although strictly outside the bounds of a direct phenomenology which attends to experience as it presents itself, there is also an indirect sense in which the sciences also "know" the temporal dimension in auditory spatiality. Given an interpretation of science as constituted by an instrumental context in which instruments extend and embody experience, the detection of microtime elements in the auditory discernment of spatial significations may also be shown. The physical understanding of sound belongs to the study of wave phenomena. Location, primarily the ability of a human listener to detect the direction of a sound source (within the limits prescribed by the experiment), may be shown to depend upon the microseconds of difference in which a given sound wave reaches first one ear and then the other. The experimenter can easily change the subject's sense of direction by varying the time elements involved, making the sound appear first on one side and then on the other side of the head or even "marching" the sound from one side to the other. However, "metaphysically," such studies present a far too complicated set of necessary reinterpretations to make them easily or readily available for a direct phenomenology. What can be tentatively accepted as an *index* for a phenomenological investigation is that by making temporal factors of the microlevel available through an instrumental context, the phenomenological uncovering of the temporal dimension of auditory experience is extended confirmedly to that microlevel.

The same degree of precision or minute analysis is called for phenomenologically. Here, in addition to the more obvious roles of repetition and duration in the temporal constitution of spatial significances, there is an important role which is made possible by the noetic operation of temporal focusing.

The phenomenon of *reverberation* as a phenomenon of auditory temporal "running off" is of particular importance as a first instance of discriminating how temporal focus is employed noetically.

It is in relation to the various types of auditory discernments of *interiors* that the temporal auditory reverberation phenomenon is most apparent. I strike a brass goblet. In its ringing reverberation I hear the resonant metallic "nature" of the object as the sounding presence of its interior. However, if I pay very close attention to the presentation of this auditory event it soon becomes clear that upon analysis not every "part" of the event is of equal importance. If I abstractly deconstruct this event as (1) the instant of the striking of the brass goblet (with my fingernail), followed by (2) its ringing reverberation *after* my finger has moved away, then it is clearly not the first tap to which I attend but to the ringing reverberation in the detecting of the interior.

There are perceptual "reasons" for this. First, in the instant of the tap there is the instant occurrence of the two "voices," the duet of things. There is the click of my fingernail on the goblet *and* the beginning of the sounding of the goblet. And in the selectivity of auditory focusing I can focus upon either one or the other; but in both cases a precise examination of the noetic act shows that the focus is upon the auditory temporal *reverberation* which follows and "runs off" the instant of the tap. The dull and almost instantaneous "failure" of reverberation of my fingernail is also revelatory of the difference of its interior "nature" contrasted to the continuance of the ringing brass. And were the instant of the tap on the brass somehow instantly damped, I would either fail to discern the interior or be seriously misled into believing that the brass goblet was something else. The instant of the tap reveals little, but the reverberation reveals a great deal. This obtains in noting that the reverberation of wood is as distinctly different from that of brass as is the distinctive sound of china or crystal from that of lead or plastic.

Thus if the "aim" of the auditory temporal focus is diagrammed according to the Husserlian model, figure 8 would be the result.

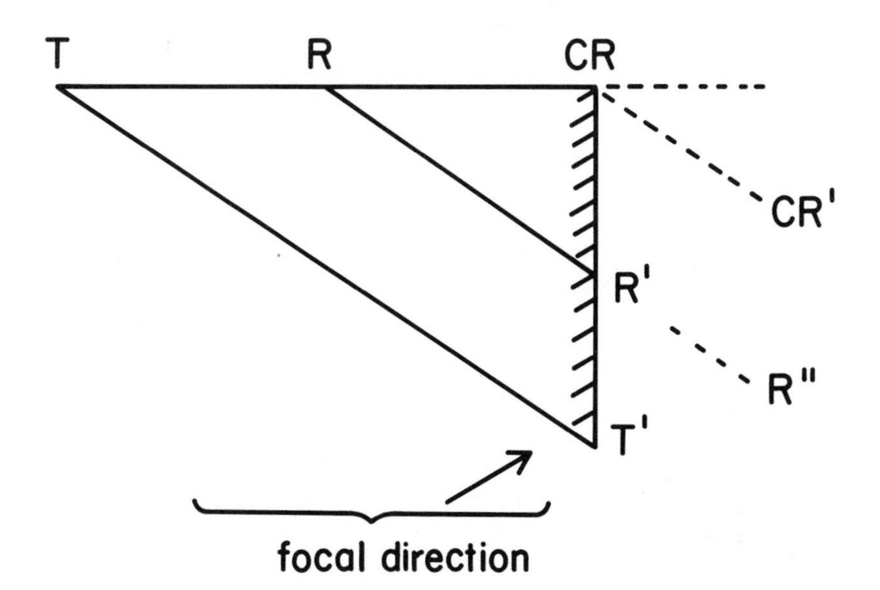

T is the instant of the tap, which is distinguishable from R, which is the succeeding reverberation. T also "trails off" however briefly in T', but it is not specifically attended to in the context of a question over interiors. R, which both continues as CR and "trails off" as R', is attended to in the act of a "fine focus" of auditory temporality.

This same following focus upon reverberational phenomena occurs in relation to interiors if the "voice" which is given is that of the *echo*. If I am practicing a human adaptation of "echo-location," as in the use of a tapping cane, or if I am blind and using this method to navigate, again I do not attend to the instant of the tap. Indeed, I do not pay much attention to the tap at all unless I am more interested in a *surface* than in the wider sense of distance and the presence of reflecting surfaces. Thus I listen for the echo, I "follow" the echo which gives me whatever sense of vague walls or corridors within which I must move. Again the "fine focus" is upon the "following" and "running off" phenomena of the whole temporal event.

Finally, the same focus upon reverberational phenomena occurs in relation to the auditory sense of distance when the question involves a giving of "voice" to things. If I call into a well, it is to the echo and its reverberation and its temporal "running off" to which I attend. It is important here that not only does the involvement of temporality in auditory experience show itself essential at every level and in the most subtle ways, but noetically there are extremely fine discriminations which point up the ultimate concrete space-time of the world.

If reverberational phenomena show themselves as important to the auditorily admittedly "weak" discriminations of shapes and surfaces and also as important to the somewhat "stronger" auditory discriminations of interiors, the noetic constitution of auditory directionality shows a different possibility of "finely focused" auditory intentionality.

I listen to the jet airplane coming into Heathrow Airport in the landing path which lies over my house. I listen with an ear to its direction. Even when closing my eyes I can "locate" it, although I note, as previously, that when I am precise and careful its auditory presence compared to its visual presence yields a slight syncopation of appearances which bespeaks a certain "accuracy" to direction if not to "location" at the distances involved. But in this instance I note that the auditory presence of the jet is a "complex" one in that "accompanying" the jet are many echoes which reverberate from the various buildings around.

I begin to perform the variations upon the many aspects of the auditory presence of the jet which show the phenomenon more fully. I quickly note that if I so desire I can focus upon the echoes which reverberate around, but the moment I do so the first clear sense of directionality in relation to the airplane recedes and becomes confused. Only when I return my focal attention to the "leading edge" of the auditory presence of the jet does the clarity of its directional location return. Contrarily, when I return to the echoes which proceed from the jet the sense of directionality becomes diminished, but I do get the usual vague senses of distant surfaces. Here the jet has become the sound source for an echo-location. But this variation begins to show that the directional location is constituted by a specific focal act which concentrates upon that utmost "leading edge" of the sound appearance. Were this example to be made strictly analogous to the previous ones dealing with shapes and surfaces, it would be to the instant of the tap that I would now attend. I auditorily seek the instant of onset (which continues) and not the "running off" which follows.

I have already noted that clicks reveal locations and directions more precisely than continuous tones. Whatever the physics of such a difference may be, here is also located a noetically discernible difference in the constitution of direction. The repeated click gives a "sharper edge" precisely to the onset of the instant than does the tone. The strictly continuous tone, particularly if electronically "pure," is "like" the stable object which "hides" its temporality and obscures its "welling up" into time awareness.

A further variation which helps locate the noetic focus at the "leading edge" of the sound event when discerning direction is the complication which occurs when the source of the sound is "shadowed." If I am listening to the sound of the jet, so long as it is overhead I can focus upon its directional presence. But when it passes over and to the other side of my house, it becomes more difficult to locate the direction and, conversely, the echo phenomena from the other buildings become more intrusive. In this instance I am positioned in an *auditory shadow*, and, as in the case of visual objects, the directionality of the sound is obscured (although not obliterated) by the intrusive object which changes the directional presence. Even in relation to my body this auditory shadow is slightly detectable. If at first I cannot get the desired preciseness of direction, I may turn my head slowly from side to side, and in the process of casting this auditory "glance" over the sounding I get the desired greater sharpness. But if I pay close attention to the sound, I may also note that the sound on the "near" side is "stronger" and "more intense" than on the other side.

These variations indicate some of the roles that temporal focusing plays in the constitution of spatial significations in hearing. It is important to note that it is by a different noetic focus that different noematic aspects manifest themselves. Thus the discernment of shapes, surfaces, and interiors, all of which are revealed through the focus upon reverberational and retentive "edges" of the temporal span, are constituted quite differently than is the case of directionality which is made precise through the noetic concentration upon the "welling up" of continuous source points or the leading "edge" of an auditory temporal occurrence.

The third possibility of temporal focus also may be taken into consideration. In the experience of music, presuming a normal and interested attention, the "filling" of space has been noted as the auditory noematic presence of music. This "filling" at peak moments surrounds, penetrates and often obliviates the ordinary sense of "inner-outer" in the musical gestalt. But this noematic presence is also correlated with the type of noetic act which does the attending. In listening to music I do not primarily attend to "things," for music provides the temptation to move away from things. Nor am I ordinarily addressing myself to practical questions concerning spaces, shapes, or even directions; and, indeed, in the presence of full orchestral sounds the problem here is one of too much rather than too little sound for such distinctions to appear clearly.

When I turn to the noetic act what shows itself reflectively in the case of this type of listening to music is that the act is not focused upon either the "leading edge" or the "trailing edge" of the reverberations of the music but is an "open" attention. Noetically the act of listening to music "spans" the full temporal duration in an "active" "letting be" of the musical presence. My protending expectations are keenly open to the oncoming "flow" of the music, and my retentional awareness is filled with the reverberations which make the music rich. Auditory focus here is "expanded" and broad (though intense). In music there is the possibility of a *field state*. This is the listening which is analogous to the visual taking in of an entire vista. It is "full" both spatially and temporally.

In the case of musical listening the "full" noematic presence of the music is correlated with the "open" noetic listening act. And at its peak occurrences in which the music "washes over and through" one in its full presence there is met the possibility of the field state in which focal attention "stretches" to the very boundaries of sound *as* present. This "stretching" and "openness" is again fully *spatial-temporal.* But this spatiality is also "thick" in that I cannot find its limits. Although I may be "immersed" in this "sphere" of sound, I cannot find its boundaries spatially. The spatial signification of a horizon is obscure. How "far" does sound extend, given some recognition of relative distance? Where is its threshold, even if I can follow a sound until it disappears? And although sounds may come from any direction how far do they "extend"? I find no clear sense of horizonal boundary such as that of the "roundness" of the visual field.

But even in the paradigmatic case of music, in spite of the tradition that music is a kind of "pure presence," I do find a sense of horizon as boundary. The musical presence does *not* extend indefinitely, although a horizon precisely in its "location" at the farthest extreme from ordinary focal attention remains difficult to discern. But in the case of listening with both its "forward" and "pastward" focal possibilities, and particularly with the "open" possibility instanced in musical listening, a sense of an auditory horizon *as a temporal boundary* does begin to show itself.

The sounds which "well up," which "suddenly appear," seem to come *from nowhere.* They present themselves continuously as having a "temporal edge" and as "trailing off" into the equal *nowhere* or nothingness of the no-longer present. It is here that we reach a clearer sense of limit characterized as a horizon, but in the case of the auditory field that horizon appears most strikingly as *temporal. Sound reveals time.*

1. J. M. Heaton, *The Eye: Phenomenology and Psychology of Function and Disorder* (London: Tavistock Publications, 1968), pp. 41–42.
2. Georg von Békésy, *Experiments in Hearing*, trans. E. G. Weaver (New York: McGraw-Hill, 1960), p. 218: "In speech there is a definite tendency to break discourse up into certain chunks and try to recognize the chunks as units, disregarding the fine structure within the chunks."
3. Edmund Husserl, *The Phenomenology of Internal Time Consciousness*, trans. James Churchill (Bloomington, Ind.: Indiana University Press, 1964), pp. 76–79.
4. Ibid., p. 48.
5. Ibid., p. 121.

CHAPTER EIGHT

AUDITORY HORIZONS

With the phenomenon of the horizon as limit simultaneously are reached the limits of phenomenology as a philosophy of experiential presence, of phenomenology under the sign of Husserl as a first phenomenology, and of the initiation of a movement to second phenomenology under the sign of Heidegger. It is also the moment when the most extreme temptation occurs to lapse into a type of "metaphysics" which would be an attempt to get beyond presentational experience by means of an "explanatory" strategy. Such a leap from description to "explanation" is the place of Democritean doubt which sense warns will be the downfall of thought.

But at the juncture where a second phenomenology begins to emerge, the strategy remains *radically descriptive*. This is the phenomenology which operates under the guidance of Heidegger and, in particular, the "later" Heidegger. It is Heidegger who has among contemporary philosophers of the phenomenological tradition most radically posed the question of a thinking which is an alternative to not merely Cartesianism but to the whole metaphysical tradition. And it has also been a hallmark of the later Heidegger to be concerned with the question of the *horizon*.

But it has also been the fate of a second phenomenology under the sign of Heidegger to be radically misunderstood, precisely because of its nearness to the poetic, its radical alterity to metaphysics, and its at first seemingly strained "descriptions" which, from the entrenched and assumed positions of a dominant metaphysics, have prevented that phenomenology from being seen as the careful *description* it is. The reading of second phenomenology entered here is one which accords to the language of a Heideggerian phenomenology its proper due as *a description of horizonal phenomena.* In the form of a Heideggerian "epoché" which "lets be" that which shows itself, it is "the things themselves" which call for the descriptions characteristic of this phenomenology of limits. A second phenomenology pushes a philosophy of presence to its final limits in the question of "absence," but it does this without reverting to the hypothesized forms of non-present presence utilized in all metaphysics.

There is a leap made by metaphysics. When the limits of sense are reached, it posits an un-sensed sense; when the limits of consciousness are reached, it posits an unconscious-consciousness; when the chain of causes threatens to proceed to infinity, it posits an uncaused-cause. And in this leap which has had such a varied history in Western philosophy as the posit becomes 'mind', 'matter', 'the Absolute', 'the unconscious', etcetera; *reality* is thought of as *other than* what is found in experience. The result is that experience invariably becomes either mere subjectivity or an epiphenomenon. Symptomatically it is the appearance of horizons which occasions this first word of metaphysics.

Phenomenology, even in its Husserlian form, begins contrarily in a "step backwards" to the roots of metaphysical origins and the origins of philosophy itself. By reviving in the most rigorous manner the science of the description of presence, Husserl placed himself at the origins of Western thought, although this did not become fully evident even to Husserl until his later writings, particularly the *Crisis.* It is there that he discovers the retrograde movement into the teleology of Western thought, going back from Descartes to Galileo, and back from Galileo to the very "invention" of the theoretical attitude in Plato. In this sense Husserl began too late in his career to open the full implications of a hermeneutic and demythologizing deconstruction of metaphysics.

But Heidegger, already having learned the lessons of first phenomenology made possible by Husserl, from the very beginning gave a historical-temporal dimension to his own version of *epoché* which becomes the deconstruction of *epochal* constitution. In *Being and Time* that historical dimension of *epoché* is the call for a "destruction of the history of ontology." And in spite of the fact that the last parts of *Being and Time* were not completed as a specific extension of that project, the deconstruction has continued to take place more and more radically in the work and thought of the "later" Heidegger.

In this context, however, what emerges in this deeper probing into the constitution of Western philosophy and thinking is the question of *horizons.* More specifically the horizon as "spatial" appears in

Heidegger's *Discourse on Thinking* as "that which regions" (*Gegnet*), and in "Time and Being," a lecture whose title reveals that the original program has not been given up, the "temporal" dimension of horizon is recognized as "Event" (*Ereignis*).[1] The matter here, though, is not a matter of interpreting Heidegger but of a certain recognition of the appropriateness of the phenomenological language of the later Heidegger precisely for the description of horizonal characteristics. For the task is to describe horizons without falling into the temptation of "metaphysics" in positing an unexperienced stratum of "experience." Second phenomenology begins with the question of the horizon as limit.

The transition to a second phenomenology is not abrupt but continuous if correctly understood. The continuity of the transition may be shown in the way the question of horizon "borders" on the previously central concerns of focal and field phenomena. The horizon as limit is an expanded ratio from the "center" of focal concerns.

This transition can be recalled from the very first appearance of horizons as limits after the first visualist model. Within the limits of presentational experience through the visual field there was a movement which began with the ordinary modes of focal or central vision toward the often implicit background and field phenomena. It should now be noted that, corresponding to ordinary focal attention and then to the question of taking note of the field as field, there is also a shift in attention and in the way in which the phenomenon is experienced (noetic correlation). The field, in turn, was seen to have a spread or expanse which eventuated in a barely noticed, vague, but nonetheless discernible border or limit. This was the first meaning of the horizon in terms of the first approximation.

From this beginning, however, a certain difficulty was also preliminarily mentioned. The "observation" of the horizon as limit is an extreme type of "observation." It is an "observation" which stands at the extreme limits of observation itself. Thus if focal attention is central and field attention is ordinarily a fringe awareness, then horizonal awareness is yet more removed from the center. Moreover, one *cannot* move the horizon into the center, and, if noted at all, the horizon is noted "from the edges" almost indirectly. Its significations are enticed only by the barely perceived motions and experiences which begin to elicit its withdrawing presence. Yet, with a certain type of questioning it becomes possible to note that it has an indefinite but roundish shape; its "edge" as limit continues to escape (direct) attention as it recedes. I fail to fix it as it shades off into the imperceptible. And I speak correctly in saying that beyond this limit is a region of the invisible, because whatever becomes visible does so only within the field of vision and must be given to that field. Outside the field lies *nothing* visible.

Within this movement and transition from core vision toward the first meaning of horizon as limit there was seen to be a set of relative relations between focal and fringe phenomena, a ratio of central to peripheral possibilities which were ordinarily structured according to a foreground and background arrangement. At the same time the essential situation of all foreground phenomena against or within a field was also noted.

Now, however, the notion of a ratio may be extended to a second level. There is a second ratio which is more expansive than the focus-fringe ratio, and it may be termed the *field-horizon ratio*. In the form of general structuring this larger ratio displays the same general features as the focus-fringe ratio. Only now we may speak of the horizon as situating the field. Roughly, *the horizon situates the field which in turn situates the thing*. This *double ratio* may be illustrated similarly to the first visualist model (fig. 9).

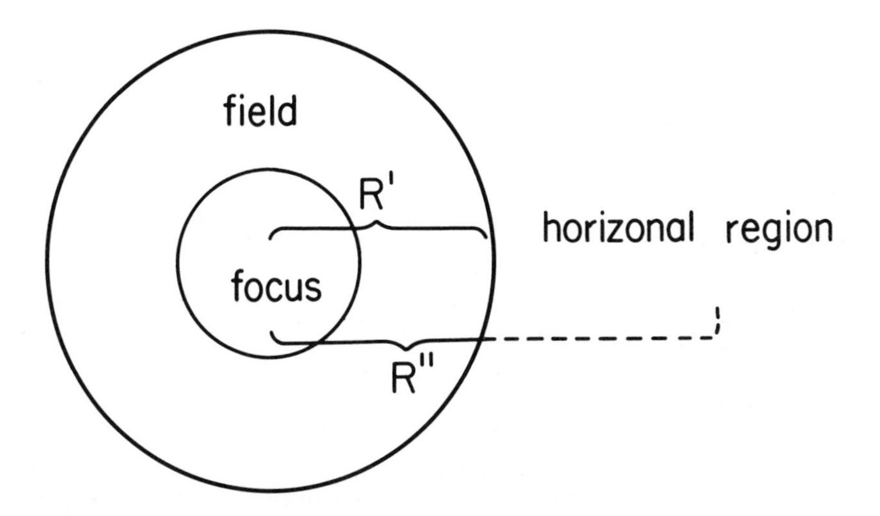

Here the first focus-fringe ratio is R', the constant relativity between focus and field in which focal objects are situated within their field. But the field itself is also situated in a constant relative ratio, R", which is that of field (including, of course, its focal center) to the horizon which shades off into the region of invisibility that is *absent* in contrast to presence. But what *can* be described is that bordering, that shading off, that *dis*appearing which does occur at the horizonal limit and the relation which that obtains between absence and presence. This is a demanding task for a descriptive philosophy of experience.

There is then a certain continuity between focus-field and horizonal characteristics in a graded ratio of extremity. If ordinary experience is normally so focally concerned that it even "forgets" the implicit field which situates the central phenomenon, so even more is the horizon as limit likely to be ignored, unattended, or "forgotten." *Ordinarily* one ignores the horizon, and even when one turns to the question of the field itself this question may remain latent and implicit. In this sense horizonal questions are "far" from the ordinary or central regions of experience and concern. By contrast, the question and noting of the horizon is in a sense *extraordinary* in what it calls for.

Yet if the horizon is that which situates the very field of experience itself as the field situates its center, then to ignore it is to risk at the least the peril of incompleteness in relation to exploring the limits of experiential phenomena. Insofar as second phenomenology makes the question of the horizon thematic, it is in line with but more radically developing the direction of thought already opened by first phenomenology. The first task for second phenomenology is to raise the question of the appearance of the horizon but without lapsing into the temptation to "leap" into that which is beyond experience in a positing and transforming of this "region" into that which it is not.

What at first seems the extreme in implicitness, once the question of a horizon is raised, yields gradually to some manifestations of horizonal phenomena with ordinary experience. This was also the case with field appearances which, once pointed out, became obvious in general outline. But with horizonal phenomena the problem remains the difficult one of locating them, since noematically horizons are the furthest remove from focal awareness, and noetically there is need for a certain "indirection" in eliciting their sense.

I return to my ordinary visual experience with an intentional involvement with the clock. It sits over there on the shelf behind me, not now appearing within my gaze. But as I turn my head the clock's presence comes into the field and I am not surprised. I have expected the "invisible" to become "visible," for in the sedimented beliefs which I have, I selectly "know" that just as the field transcends the thing, so the World transcends my opening to it. To name the horizon in relation to the visual field is to name the World. But in the very way in which I take the World for granted I may miss the phenomenon of horizon.

This first location of horizons as the place within experience, where the enigmas of withdrawing and absence occur as the "signs" of "transcendence" appear, has until now been expressed in terms of the borrowed clarity of a visualist approximation. The horizon of the visual field shows itself most easily as a spatial signification, a border or limit in its first appearance. And although vague, it has a shaded-off shape which is roundish.

With the auditory turn, however, the horizon does not show even such vague spatial significations except as thresholds of hearing. We are so situated "inside" the auditory field that its indefinite extent is not detected primarily in spatiality. It may be "spherical" in the sense that this "opening" extends indefinitely outwards such that we are "surrounded by" the fullness of auditory spatiality. But in the "strength" of sound a horizonal presence is discernible *temporally*. It is indeed the place where the auditory dimension of horizon is most dramatic.

As I listen to music on the radio, the notes "well up" out of the "nothingness" of the future and "trail off" into the horizonally equal "nothingness" of the past, and the sense of horizonal "absence" is the experienced temporality of sound. These sounds "give themselves" into presence and then "fade out" in the temporal dance of the auditory dimension.

In the previously located ability to shift the auditory temporal focus so that listening may search out the "leading edge" of sounds just coming into presence or follow the sounds as they "trail off" into the reverberations of the just-past of retentions, I also vaguely and indirectly detect the temporal horizonal limit of auditory experience. My expectant protentions and my most intense retentions always "break off" at horizonal limits.

In such variations which gradually elicit the sense of the horizon, I note that the general characteristics of the difficulty and the implicitness of horizonal discernment remain isomorphic with those characteristics noted in visual experience. The horizon is that most extreme and implicit fringe of experience which stands in constant ratio to the "easy presence" of central focusing. There is also a *resistance* offered by the horizon. It continually recedes from me, and if I seek for sounds and the voices of things, I cannot force them into presence in the way in which I may fix them within the region of central presence. I must *await* their coming, for sounds are *given*. But when they are given they penetrate my awareness such that if I wish to escape them I must retreat "into myself" by psychically attempting to "close them out."

Here the attempt to describe horizonal phenomena comes upon a language already partly Heideggerian. To *describe* the horizon calls for such language, for it is a correct description of the phenomenon of the horizon. Once located, the horizon as that limit to experiential presence may be described at its borders. There the horizon is a *receding*, a *withdrawing*, that which is *beyond* what is in presence. The horizon is the limit where presence is "limited" by *absence*. The horizon continually withdraws so that its *open* "region" is itself never present. A second phenomenology brings up the question of absence.

All of the above are *descriptive* of the horizon as a limit. But if now the horizon is located as such a limit, and it is understood that the double ratio of focus to field and field to horizon obtains, then the absent, withdrawing region may be viewed as that openness which situates the whole of experiential presence. What is present is always found "inside" or *within* horizons; what is "not given" locates what is given. But although this borders on metaphysics, it is not metaphysics; at least so long as the open horizon is not made into what it is not: a spurious form of presence. "In itself" the horizon continues to be *hidden*, to withdraw, to present itself as the open region. But even such a recognition of the non-presentability of horizon as it "is" beyond, in relation to the double ratio, allows one to characterize what is present, the field of experience, as being in an *abiding expanse within which things are gathered.*

Within the auditory dimension one may add the characteristics of horizons in terms of temporality. In the lecture, "Time and Being," Heidegger speaks of an *Event* as a *giving*. Being, which is that which comes-into-presence, that which is (already) gathered, is the *given*. But at the horizon one may note the giving, the *e-venting*, the point at which "there is given" into what is present.[2] Nowhere is this more descriptive than in the experience of listening. The sounds "are given," they come unbidden into presence, and humankind, in listening, is let in on this e-venting. Listening "lets be," lets come into presence the unbidden giving of sound. In listening humankind belongs within the event. And as a presence, the sound is that which *endures*, which is *brought to* pass, the sound *whiles away* in the temporal presencing which is essential to it.

Nowhere has this phenomenological language been more descriptive of the phenomena, nowhere more literal, and nowhere more Heideggerian. Yet in relation to horizonal phenomena this is proper description. Presence is situated within its horizons, and at the extremes of horizonal limits can be discerned the "coming-into-being" out of the open and absent giving (*Ereignis*) and the region (*Gegnet*) which is "beyond" presence.

But in pushing horizonal description to its own limits, one further identification is called for. How is the horizonal "absence" which "sends" that which is received to be auditorily and temporally characterized insofar as it is heard? The clue lies in the enigma of *silence*. Silence is a *dimension* of the horizon.

The enigma of silence is in how it is "given in absence." Here the full enigmas of language necessarily meet the enigma of experience faced with the question of horizons. What "is" in the language which describes presence is sound rather than silence. And until the question of a horizon is raised, it would be quite possible to fail to discover silence. Experientially, I cannot escape sound. If I return eidetically to the field presence of sound it is continuous.

There remains, however, a sense in which silence is "given in absence," and its withdrawing horizonal absence may be detected in the most mundane of experiences with things. Silence belongs to those mute objects which have pursued philosophy through the ages. Silence belongs to the syncopation of experiences in which what is seen seems silent while what is not seen may sound. In this one could almost say that silence is a "visual category." The pen on my desk, the vase on the mantle, the tree now still in the absence of a breeze, lie before me in silence, until echo or contact awakens a sound.

This is a *relative* silence. Silence adheres to things hidden relatively within present experience. In Husserlian terms, silence belongs to the "empty intention," the aim of intentionality which is co-present in every intention but which is the "infinite" side of intentionality that does not find fulfillment. There is a "depth" to things which is revealed secretly in all ordinary experience, but which often remains covered over in the ease with which we take something for granted.

For a brief moment, returning to visualism, this "depth" may be noted in the phenomenon of perspective. When I view a thing it presents itself to me with a *face*. A deeper and more careful analysis reveals that it is not just a *surface* face, but a face which is an appearance that presents itself as "having a back" as well. The "illusion" of depth which is possible for stage settings and two-dimensional pictures is possible only on the basis of a "real" depth which belongs to naïve existential experience. The thing "transcends" its present face in its "absent" but intended back. The present is bounded by what is meant, or intended, which is the *adherence of horizon to the hidden side of the things*.

In this sense the thing may also be spoken of as having a horizon. Its hidden depth, its absent profiles which, while they may be given, are always those which recede when the thing is presented before me. And, here, too, the descriptions of horizonal phenomena continue to hold true. The thing displays a constant ratio of present to hidden, of visible to invisible.

Auditorily this hidden depth is silence. In its relative horizonal features silence lies hidden along with the sounding which presents itself. But silence, as in all horizonal features, is not a matter of contrast or of opposition as such. Silence occurs in adumbrations of the soundful to the silent. Even prephenomenological understandings of listening have their versions of this "shading off" of sound toward the silent. Just as the visual is co-presented as a face in profile and an intended absent depth, so in listening there is a sound which shades off into the co-presenting "emptiness" of silence. Silence is the "other side" of sound. Relative "absences" of sound have often enough been understood to belong properly to "meaningful" auditory experience. The pauses, or rests, in musical phrasing add to rather than subtract from the totality of the music. In speech silence often indicates either the stopping of a line of thought or a transition, but silences can also be filled with their own significations.

There is, moreover, an aspect of intentionality which "gestures toward" silence. In listening to music, particularly reproduced music, there is the experience of the *intrusion* of unwanted sound. The hiss, hum, and static which may occur distracts, and one "gestures toward" the silence which allows the music its "purer" presence. Ideally, if music is to reach its full presence, it must be "surrounded" or "secured" by a silence which allows the sound to sound forth musically. This is one of the aims of a set of headphones which do not so much improve the music as help close out the other sounds and thus procure a relatively "surrounding" silence.

In the adumbrations of sound toward the silent there is also a relative silence which is "filled" with signification. In conversation when the other is silent there is also a "speaking": we see the face which "speaks" in its silence. We feel the flesh which "speaks" in its silence. There is an adherence of speech to the silence of the other.

Such adherences within relative silence enrich with auditory depth things and others. Even mute things may "speak" in a silence which carries the adumbrated adherence of sound to presence. I look at the postcard which arrived recently from Japan. It depicts four peasants running from a sudden rainstorm. They hunch under grass hats and mats as they seek shelter from the wet coldness of the rain. And if I look intently at the picture, perhaps mindful of the dictates of a Zen passage read long ago, I detect the adherence of a certain auditory presence to the picture. I "hear" the rain and "listen" to the peasants running and to the rustling of the mats. The muteness of the picture "sounds" in its relative silence.

But it is not as if "silence itself" were discovered. The silence of horizonal phenomena continues to withdraw, but in its withdrawing may be heard the *giving*, the eventing which sound is in its coming-into-presence. *Beyond* this limit silence continues to escape. Heidegger has characterized the horizon in this respect as a *Nearness* which in its near distance has at once the character of refusing and withholding. He also called it "the hidden nature of truth".[3] It is revealed only in the withdrawing, "the horizon is but the side of that-which-regions turned toward our re-presenting. That-which-regions surrounds us and reveals itself to us as the horizon."[4] And here are also reached the limits of a "direct" description in the enigma of giving word to this horizonal openness. "Any description would reify it [that-which-regions] . . . nevertheless it lets itself be named and in being named it can be thought about . . . only if the thinking is no longer a re-presenting."[5] Such a thinking as is appropriate to the horizon which shades off into its "absolute" beyondness can only be characterized as a thinking which is a *waiting*. Waiting is the limit of all "empty" intentionalities. Waiting is a "letting be" which allows that which continuously "is given" into space and time to be noted. Auditorily this is a listening to silence which surrounds sound. "Silence is the sound of time passing."[6]

1. I shall utilize William Richardson's unpublished translation of *Ereignis* as "event" rather than the published version of Joan Stambaugh who renders it "appropriation." Martin Heidegger, *Time and Being*, trans. Joan Stambaugh (New York: Harper and Row, 1974).
2. My exposition continues to rely on the unpublished Richardson translation.
3. Martin Heidegger, *Discourse on Thinking*, trans. John M. Anderson and E. Hans Freund (New York: Harper and Row, 1966), p. 83.
4. Ibid., pp. 72–73.
5. Ibid., p. 67.
6. Thomas Stoppard, "The Bridge."

PART THREE

THE IMAGINATIVE MODE

CHAPTER NINE

THE POLYPHONY OF EXPERIENCE

The first movement of a phenomenology of sound and listening has taken its first step in what may be regarded as a preliminary survey of the auditory terrain. It began with first approximations and the center of focal listening. It moved from that listening to the voices of things "outwards" and from there to the listening for the silence of the relative and open horizon of silence. This survey has been attentive to the voices of the World.

This is phenomenologically appropriate, for there is a *primary listening* which precedes our own speech. This is whether one considers the matter as an issue of personal history—I hear the voices of others, of things, of the World long before I speak my own words—or as a matter of the correct phenomenological procedure which begins with noema before taking up noetic acts. Phenomenologically the "self" is modeled after the World which takes primacy in its first appearance.

The movement toward a more detailed review of the auditory terrain is a movement which accelerates the approximations to *existential significations*. The sounds which we hear are not "mere" sounds or "abstract" sounds but are significant sounds. In the first instance listening is a listening to *voices*, the voices of language in its broadest sense.

Existentially things "speak." Heidegger has pointed out, "Much closer to us than all sensations are the things themselves. We hear the door shut in the house and never hear acoustical sensations or even mere sounds. In order to hear a bare sound we have to listen away from things, divert our ear from them, i.e., listen abstractly."[1] The things of the world sound in their own way. Things, others, the gods, each have their *voices* to which we may listen. Within auditory experience there is this *primacy of listening*.

Not only do things, others, the gods, and ultimately I "speak" in distinctive voices, but each has its own way with language. For within auditory experience I find myself already *within* language. It is already there. Existentially there is already "word" in the sounding wind which brings things, others, and the gods to me. There is a sense in which within experience a "prelinguistic" level of experience is not to be found. The "prelinguistic" is the philosophical counterpart to the "preperceptual" bare sensation which if found at all is found by diverting one's ears and eyes from the objects.

Sounds which are heard as *already meaningful* do not show us the "lost" beginning. The actual history of man who speaks before he learns to embody word differently in writing and in wordless symbols does not show us the hidden genesis of the word. Nor do prescientific and preliterate languages show us the beginning. Not even the child's "learning" of his first word which contains in itself however latently the "whole" of language reveals the genesis. Long before he has learned to speak he has heard and entered the conversation which is humankind. He has been immersed in the voices and movements which preceded his speaking even more deeply in the invisible language of touch and even that of sound within the womb. Listening comes before speaking, and wherever it is sought the most primitive word of sounding language has already occurred.

The presence of word *already there* for listening is also what I find if I inquire into myself. For wherever I find myself I already stand in the midst of word. My memories do not give me that "first word" which I uttered as a child nor even the "first word" I heard from my parents. This lies beyond the horizon of my memory and appears if at all as an already mythical tale related to me by others.

Nothing gives me the "lost" beginning of word spoken by voice either as that which is built up or as that which occurs at a stroke. Nor is there a need for phenomenology to search for such a beginning. If I listen, I may begin in the midst of word for there is a *center* to my experience of language. It is that strange familiarity which lies in the very conversation which shows things, others, the gods, and myself. *The center of language is located in the voiced and heard sounding of word.* And from this center I may proceed "outwards" toward the horizons of sound and meaning which embody significance within the World.

The voices of the World find response, an "echo," in my own voice which takes up the languages of the World. My "self" is a correlate of the World, and its way of *being-in* that World is a way filled with voice and language. Moreover, this being in the midst of word is such that it permeates the most hidden recesses of my "self." It is for this reason that the more detailed review of the auditory terrain which follows not only moves ever closer to existential significations, it also takes note of a modality of experience so far barely noticed in the first listening to the voices of the World.

To this point the themes which have been followed could be characterized as *monophonic*. It is as if I, the listener, have been primarily a "receiver" of the voices of the World. As experiencer I have not yet spoken, nor have I yet heard all there is to hear. In particular I have not yet paid attention to that second modality of ongoing experience, the *imaginative mode*. With the introduction of a second modality of experience, in addition to what has been the predominantly perceptualist emphasis, listening becomes *polyphonic*. I hear not only the voices of the World, in some sense I "hear" myself or from myself. There is in polyphony a duet of voices in the doubled modalities of perceptual and imaginative modes. A new review of the field of possible auditory experience is called for in which attention would be focused upon the co-presence of the imaginative.

If the first survey weighted perception, it did so in terms of what has been taken as primary in first phenomenology. Yet even within first phenomenology there is also a counter tendency. Husserl elevated imagination as fantasy to the level of the privileged "instrument" for critical phenomenological reflection itself.[2] His paradigm, at least implicitly, was the thinker such as the logician or mathematician who could reconstruct whole "worlds" by himself. But this elevation of the imagination could equally have been properly modeled after literary or poetic or artistic thinking which also surveys the possible.

Husserl's use of imagination, moreover, often revolved around the reproductive or representational capacities of imagination. Here imagination reproduces what was previously perceptual, but in the assumption that it does reproduce the other modes of experience lies a threat to the primacy of the perceptual itself. In imagination, even at the level of variations, there is already an "excess" which carries it beyond perception. Hidden in this "excess" are both certain aspects of "self-presence" and of a fundamental liaison with the World. The "innermost" is *not* distant from the "outermost."

The imaginative mode, to be considered very broadly as ranging from the "empty" supposings to the most concrete "images" of thought, contains within itself the variations of "self-presence" and "thinking" which pose such difficult questions to philosophy. Of course here the primary question of context is one directed to the various forms of *auditory imagination.* What is it that I "hear" when my listening is to the "second voice," the imaginative voice? What is that listening which occurs within my self-presence and which accompanies the presence of the things and of others in the perceived World? If the "self" arises phenomenologically strictly as a correlate to the World, a correlate which is in a sense discovered only after discovering the World, it also hides within itself and its imaginative acts (which hide themselves from others) a kind of autonomy.

In the auditory dimension the imaginative mode is a matter of "voice" in some sense. Its center, suggested above, may be located in a clue provided in the history of phenomenology by Merleau-Ponty.

In the discussion of the body as expressive, Merleau-Ponty notes that what is usually taken as an inner silence is in fact "filled with words" in the form of what will here be characterized as "inner speech."[3] *Focally, a central form of auditory imagination is thinking as and in a language.* With and around this phenomenon revolve many of the issues which relate to specific human experience and self-experience.

The second survey which begins its investigation of the polyphony of experience binds what is "innermost," the imaginative, with what is also the broadest in human experience, the *intersubjective.* It is the *voices of language* which assume a focal role in human imagination in its auditory dimension.

Initially there is nothing more "obvious" than the familiarity of human speaking and listening. Wherever humankind is found it is found speaking. Through the polymorphic shaping of sound sing innumerable languages. Languages bind together and separate humankind. Otherness and strangeness is dramatic in the difference of tongues, but there is also the human ability to learn to "sing" in any language.

Language also lies in the interior. Inner speech as the hidden monologue of thinking-in-a-language accompanies the daily activities of humans even when they are not speaking to each other. The voices of others whom I hear immerse me in a language which has already penetrated my innermost being in that I "hear" the speech which I stand within. The other and myself are co-implicated in the presence of sounding word.

Phenomenologically I already always stand in this center. The voices of language surround me wherever I turn, and I cannot escape the immersion in language. The voices of language have already penetrated all my experience, and this experience is already always "intersubjective." And if this experience of the omnipresence of language which comes from others and which settles even into the recesses of myself is "like" the experience of surrounding, penetrating, pervasive sound, it is because its ordinary embodiment lies in the listening and speaking which embodies the voices of language. Voice is the spirit of language.

But if the voices of language are the central theme for the polyphony of experience, a survey of the wider reaches of the imaginative mode is also needed. Before the dominant feature of the "inner landscape" can be determined, a series of approximations which more clearly locate it is called for. The voice belongs to a vaster polyphony of perceptual and imaginative experience.

To begin with the ordinary, as I turn to "inner" experience in the mode of the imaginary, I note that these experiences may "echo," "mimic," or "re-present" any "outer" experience. Imagination *presentifies* "external" experience. I *see* the butterfly alight upon the sweet pea; I close my eyes and recollectively imagine the same event. I heard the distant foghorn in Port Jefferson, but I can imaginatively remember it now. These re-presentations may be exceedingly varied in form as memories, recollections, or fantasies, etc., but in each of these they display themselves as *irreal.* It is not that irreality is lacking in vividness; a lack of vividness may be a contingency of a particular person's imaginative ability, or it may be the result of a lack of attention and "training" in imaginative acts. But the irreal presence is marked by "immanence" as "mine" and as "hidden" from the other.

But if the irreality of the imaginative contrasts with the sense of actuality and transcendence displayed by "outer" experience, there remain many respects in which imagination displays a structural isomorphism with perception. Imagination, like perception, is susceptible to further and further refinements, discriminations, and enrichments as the rich imaginations of artists have revealed through the centuries. Auditorily Beethoven was able to imaginatively "hear" an entire symphony at will. Even after deafness his "inner hearing" did not fail him as the magnificent Ninth Symphony so well shows.

With this variability and polymorphic capacity for refinement possible in imaginative modes of the experience, the dangers to a descriptive phenomenology are encountered again in the temptation to arrive too soon at a superficial, if apodictic, level of discovery. The richness of imagination is at least as complex as that of perception.

I return to preliminary imaginings. If various forms of the irreal may re-present perceptual contents in the form of memories and recollections or fantasies, there also occurs in each of these forms the "inner" capacity to *vary presences indefinitely*. Usually this capacity of imagination is centrally located in fantasy, but the irreal presentifications in each of the forms of imaginative activity may occur *either spontaneously or at will.*

I may remember the sounds of the workmen arriving with their usual clatter yesterday in the mode of a mnemonic repetition of that event, and this remembering may occur "at will" or "spontaneously." Or, if I am imagining as a type of fantasy, I may lie back daydreaming, allowing my "thoughts" to drift before me. Equally, if I am searching out a problem, I may, in a disciplined exercise of variations, try at will this and then that alternative.

Between spontaneity and at-will presentifications lie other gradations of possibilities. The occurrence of one imaginative content may spark by "association" a series of others, or a "line of thought" may lead "linearly" like a deduction to something else, and so forth. But in the midst of at-will imagining, particularly in fantasy, further possibilities of the "inner" modality show themselves. In fantasy the variations are not merely irreal, they are "free" of the intentional re-presentations which mark recalled and mnemonic occurrences. Thus I can imagine centaurs, satyrs, creatures from Mars, or the catalogue of imaginary beings classified by Borges. This fantasizing applies to auditory imagination as well. I may "hear" the stellar music of *2001* or turn the enchanting songs of humpback whales into a chorus at will.

The range of variability of "inner experience" is as wide and as susceptible to learning as that of "outer experience." But in some respects there is an "excess" of imagination over perception. Imagination is not a mere mimicry of the perceptual. This is not to say that imagination is absolutely free, because, as previous phenomenologies of the imagination have shown, imagination like perception has its own distinctive structures and possibilities.

Imaginative acts also implicate the "self." As "my" imaginings, particularly those which I presentify to myself at will, the sense of an "inner self-presence" entices the very notion of a "self." In imagination I am able to "experience" myself. But the way imagination "shows" a "self" may vary considerably. In class, concerned with showing existential possibilities of imagination, I ask the students to imagine experiences which they have not in fact actually experienced before and to describe what they imagine. One student imagined himself jumping from an airplane in a parachute, an experience he had never had but desired to have. Upon more specific inquiry we discern that he imaginatively "feels" the rush of air upon his face; he "sees" the ground rushing up to meet him; he "hears" the airplane receding in the distance. Imaginatively the full play of the "senses" is vividly presentified.

A second student, however, describes the same type of experience very differently. He "sees" himself jump from the airplane. He does not "feel" the wind or "see" the rushing of the ground to meet him but "sees" himself "out there" as a "quasi other" jumping and falling toward the ground. Upon repeating these exercises in different classes, this difference consistently emerges. "Empirically" some self-imaginations are experienced as occurring "in" and "from" one's own body, while others are *objectified* in that they place themselves "out there" apart from their sense of body as an "objectified quasi other" in the imaginative experience.

But once this difference is clarified and compared, most of the students find that they can vary embodied and objectified self-imaginations at will. However, it remains so that in the objectified mode it is almost always the case that the "quasi other" who is my "self" is not only "apart" from the sense of "being in" one's body but is displayed without the full range of sensory imaginative presence. There is a lack of a "feeling in" my "self" as "quasi other."

Again, these imaginative possibilities as spontaneous or at will, as embodied or objectified, are also locatable in auditory imagination. Spontaneously, though in the mode of irreal imaginative recollection, the brilliant passage from the harpsichord solo heard last night in the Purcell Room may occur to me. Or, if specifically attending to this occasion, I may at will try to remember the continuo of the cello in the trio sonata which opened the concert. In auditory fancy I may also at will imagine a set of disharmonic tones and in fact either build them up "one by one" or "hear" them occur in a gestalt.

Embodied and objectified auditory imaginations may also occur. One dramatic and sometimes "pathological" phenomenon are the well known disembodied "voices" heard especially by schizophrenics. "Voices" occur spontaneously, and sometimes the patient is not even able to tell if it is his own voice or that of an other; whether it is from within or from elsewhere. But far less extremely, it is possible to "disembody" one's own voice in auditory imagination and hear it as from a tape recorder. Here in an imaginative version of an auditory mirror are elicited a series of difficult problems which revolve around imaginative self-presence. What, then, is the form of an embodied auditory imagination? Do I, whenever I turn to "hearing myself" speak, objectify my voice as that of a "quasi other"? Or does there lie so close to "me" a most familiar and thus most difficult to elucidate embodied auditory imagination which is the ongoing presence of a dimension of my own thinking, an "inner speech"? Although such a suggestion is almost too easy and too obvious, its location calls for yet further approximations.

Do, for example, the structures of focus, field, horizon, spatial, and temporal features appear in the imaginative mode, and if so in what form? As one enters further and further into the "observations" called for, more and more difficulties are encountered. These difficulties, however, are not ones that are to be blamed on the "introspective" nature of imagination rather than the "extrospective" nature of perception nor because imagination cannot be checked "publicly." In that respect imaginative contents are no more enigmatic than perceptual ones, for I no more have the other's perceptual hold on the World than his imaginative one. And in other respects he can as easily report his imaginative activity to me as he can his perceptions. Phenomenologically there is as much intersubjective validity to the exploration of imagination as there is to perception, and in both one must first seek "for himself."

What does make the "observation" of imaginative phenomena difficult is the very essential variability and ease of presentification which belongs to the noematic presence of the imaginative content. It is of the very essence of the imaginative noema to be easily changeable and variable. For no sooner do I "think of it" than it is "there." Its dissolubility, its rapidity of transformation, its vivid but "evanescent" presences make it difficult to "fix" what is imagined.

This "flux" of appearances, the apparent "insubstantiality" of them, the "flow" of them as events have shown themselves before. This characterization of imagination is "like" the first characterization of the *auditory dimension*. Moreover, the "flux" and "flow" of these features implicate again the sense of temporality which belonged dramatically to listening. As in the first exploration of the auditory dimension, one suspects that these features belong as much to the *first stage of reflection* as to the phenomena. The Heraclitan dynamism of the imagination may be but the preliminary appearance of its being, but there may also be a *secret liaison* between the "flow" of imagination and the "flow" of the auditory. Each begins in the same grammar.

Given this initial difficulty regarding imaginative presences and a suspicion about what occurs at a given level of reflection, the question of structural features becomes a matter of more subtle and careful variations in imaginative intuiting. There is also to be preliminarily considered the same problem facing an investigation into the imaginary as previously discerned in relation to the perceptual. In the traditions regarding the imagination the "image" has become as discrete and "atomized" and isolated as the infamous "sense datum" of perceptual experience. But in imaginative experience there is an even more difficult problem again located in the presence of the imaginative noema. This difficulty emerges in relation to the question of a focus-field relation in imagination.

I imagine a philosopher's Pegasus, varying his color—now purple, now white, now green—and varying his shape—now with ragged wings, now with smooth wings. In each variation he may appear vividly in the mode of irreality. There is no doubt that the variable Pegasus "stands out" in imagination. But no sooner do I raise the question of a field than a difficulty emerges. For if there is a field it "shows itself" not only as implicit but so obscure and indefinite and unstructured that I begin to wonder if I presentify a field upon demand. Yet, when I return to further and more extreme variations, I do begin to detect the sense of a field.

I imagine a transparent Pegasus which I now "see," and here I begin to detect that he is etched out against a background of almost indiscernible color and extension, which is a field nevertheless. But *unlike* the focus to field relationship in perception, the starkness of contrast between the core figure and the vagueness, indefiniteness, and much higher degree of implicitness, the contrast is so marked that without careful variations it is possible to almost miss the field. Of course one existential possibility of a field may be detected immediately upon raising the question, for I can imaginatively "supply" a vivid field for my Pegasus and make him in his purple variation stand out against a green ground. But by not attending to background, the imaginative field remains much more starkly indefinite and much further a fringe phenomenon. Here we come upon a difference between the perceptual and the imaginative modalities. There is a higher degree of contrast possible between focus and field in imagination than in perception.

But if the contrast between focus and field is a matter of severe degree, there is another respect in which an even greater contrast occurs. In perception, regardless of how fringe-like the presences are, there is a *constancy* of perceptual presence so long as one is experiencing at all. But in imagination a whole dimension of imaginative experience may be "turned off" and absent. Neither focal figure nor field background can be detected. This is most easily noticed in relation to the more often neglected dimensions such as those of smells. It is quite certain that I find no ongoing imaginative smell signification except as bidden or as may occasionally occur spontaneously. In fact some persons are quite surprised that they can imagine smells, and this may at first be difficult. It should also be noted in passing that there are also cases in which a person is lacking entirely some imaginative dimension such as that of visual imagination in which case there is an "imaginative blindness."[4] Apart from these cases, the fully dimensioned imaginer finds it quite simple to "turn off" one or another dimension of imaginative experience. Speculatively, this may be one source of the temptation to "atomize" and make discrete global sensory experience. The imagination in its variability which "exceeds" that of perception "allows" this possibility. In this aspect imagination, *although not perception*, is latently "analytic."

If an entire dimension of imaginative experience may be "turned off," when it is "on" it continues to display itself as "like," its perceptual modality though different in its degree of internal relations between focal and field aspects. Colors, shapes, extendedness, figure and field, even three dimensionality may be noted in visual imaginations. But there is one interesting respect in which the imaginative visual field is *not* isomorphic with its perceptual modality. In regard to field shape the imaginative visual field "exceeds" its perceptual modality.

I imagine a green bee buzzing "before me." At will I gradually move him to the side, then to the *back* of my head; yet I still "see" him in imagination. Here, however, is reached one of the most difficult to determine imaginations. I wonder about this possibility. Have I subtly deluded myself and disembodied myself? This variation is indeed easy to perform: I imagine myself sitting in a chair with the green bee buzzing behind me, but I am now "seen" as a "quasi other." This is distinctly different from the imagination of the green bee "behind" me in the embodied form of imagination; yet he presents himself here, too. I continue to "see" the bee behind me, although I have to admit that his imaginative presence does change. He is now no longer only a "visual" bee, he is also "felt" and "heard" as in the cases of the adherence of significance beyond the limits of a horizon. He also is "visually" imagined. In this, the shape and limit of the imaginative *field* "exceeds" its visual limitation of being "before" one. But in the imaginative modality this "excess" is one which has already previously shown itself in other field shapes including that of hearing. In a "likeness" to an auditory field an imaginative "visual" field is omnidirectional. In imagination the field-shape possibilities of the visual dimension are closer to those of an auditory field-shape than in the perceptual mode.

But in the question of the dissimilarity of imagination and perception, the "turning off" at will of what is for perception a constancy, raises another line of inquiry. Is such obliteration of entire imaginative dimensions a relative or an absolute possibility, or is there *continuously* some form of imaginative self-presence? Here a further descent into the imaginative mode is called for.

This question whether it is possible to completely "turn off" the "thinking self" insofar as it is imaginative is not without purpose. But its purpose is not to answer the absolute question, for if this were possible then the answer might be one which must face the problem of cessation of experience. It is rather an indirect way of eliciting the modal possibilities for locating the continuities which show themselves as "contingently" dominant in the "thinking self." For the suspicion is that at least so long as one has awareness, *some* modality of imaginative "self-presence" occurs. It is in this line of inquiry that further approximations to a polyphony of experience may be noted.

To more precisely localize and isolate such dominant continuities another set of graded reflections must be addressed to the polymorphy of global experience in both its perceptual and imaginative modalities. The first task is one of detecting the *co-presence* of perceptual and imaginative modalities.

Simple variations show this quickly but in such a way that there is again a graded set of what is focal and what is fringe within the multiplicities of global experience. That is, globally anything that is present to any degree will be placed within a stratification which is gradated from a "center" of attending focus to an extreme fringe, "vague" presence. This gradation and ratio show themselves in the simple variations which locate the existential possibilities of perceptual and imaginative co-presence.

I am sitting in my chair, deeply engaged in thought, perhaps specifically going over a set of imaginative variations about a certain problem. The perceptual awareness of my surroundings as a whole is "drawn in" and becomes fringelike in character. My usual bodily sense of fringe "weightiness" and "locatedness" gradually transforms into an almost "weightless" feeling, and my feet, propped up on the footstool, almost seem to "disappear." I am so deeply absorbed in my thinking that my bodily experience and the surroundings are "almost forgotten." The "inner attention" is the focus of global experience in contrast to the now implicit "outer" fringe. But suddenly Lisa bursts into the room to announce her latest success at school. I jump, and with an almost instantaneous "switching" of focus I find my global experience "outwardly" directed.

In the case described above, however, the polymorphy of imaginative "thinking" co-presence is detected. In ordinary situations it shows itself as "weighted" toward either an "inner" or an "outer" focus, a ratio of focal-to-fringe awareness. When I am engaged in an activity, particularly a demanding bodily activity, the converse of the absorption in thought occurs. I chop wood for the evening fire in Vermont. Here I am "in" the ax as my energies and forces are directed through the wood. I am "outside" my "inner" self, not to be distracted for the very obvious reason that if my attention was not so concentrated, I may well lop off my foot. Here, while there is also an "inner" awareness, it is fringelike with respect to the active embodiment in the chopping.

A third variation begins to show more balance in co-presence. I sit down to read, the book lies before me "perceptually," and I must "see" *in* the words the thought that they may bring me. Of course it is not the "shape" or the "surface" of the letters which I attend to, but the "meanings" in the words. But while this activity proceeds I may find myself "wandering off" into my own thought, perhaps stimulated by what I have read, and before long the reading has receded and become "mechanical" as I move from the meaning in the words to focus upon my own "inner" meaning thought. One or the other side of imaginative-perceptual co-presence ordinarily takes precedence. Such variations establish the essential possibility of co-presence but do not exhaust it.

I press the investigation further. If it is possible for two modalities of experience, perception and imagination, to be variably co-present, is there any structure more detailed which allows this co-presence to be heightened or lessened? The new line of inquiry leads into considerations of the dimensional characteristics of imaginative-perceptual co-presence. I begin to note a series of *resistances* within the possibilities of general ratios of focal-to-fringe experiences.

I look at the picture of the sailing yacht on the wall before me. All the while I am thinking of how much I desire to have such a yacht, but I also tell myself that this is a luxurious dream. Or I gaze almost blankly at the pile of correspondence on the desk while thinking of the agenda for tomorrow's meeting. Here, still within the general structure of a variability of focus and fringe, there is a ratio of co-presence, for there is some awareness of both the thought and the perceptual presence. There is a "distance" between what is imagined and what is perceived. But the "distance" is more than a "distance" of imagination and perceptual modalities, it is also a "distance" of dimensional aspects. I see the yacht, but I "think" in "inner speech" what it is I think about the yacht.

If now I look at the yacht and try at the same time to visually imagine it, I find a *resistance* which shows itself in several ways. First, I find that in the attempt there may be a subtle but detectable *alterna-*

tion between the imagined and the perceived yacht. Within the *same* dimension of experience there is a conflict between imagination and perception; there is the need for a kind of "distance" for the co-presence to be easy and distinct. The "distance" is a matter of a *harmony* of co-presences within polymorphy.

Such harmonies of the "inner" and the "outer" modalities is in fact quite ordinary. The commuter driving to work can be quite aware of the habitually experienced flow of traffic while at the same time intent upon his plans for the day. So long as there is a "distance" there may be co-presence. But within the same dimension the loss of that "distance" produces a resistance which may also take several forms. There is the already noted resistance to perceiving and imagining the same thing at the same time. There is also an occurrence which is frequent, but which also often escapes explicit notice. That is a *synthesis* of imaginative and perceptual co-presence in the form of a very ordinary "hallucination."

This "hallucination" is one in which a particular type of co-presence momentarily synthesizes such that in that moment what is "imagined" is "seen." If I am intent upon my reading, the ideas flow smoothly into my awareness, and I pay little specific attention to the words as such. But if my thinking begins to wander and gradually drifts away from its previous concentration, the reading gradually recedes but remains present. If I realize what is happening I may immediately return to the reading, but sometimes in the drifting of thought I may think along a similar or related line of thought and suddenly may "see-imagine" what appears actually to be a "wrong" word. *Crescent* may be what is written, but I "see" it as the word *present*, as if in a "perfect" but momentary synthesis this transformation had occurred.

A moment's reflection and rechecking shows that the synthesis was momentary, but here the polyphony of the co-present modalities of experience "blended" so that the appearance was neither a "pure" imagination nor a "pure" perception.

This momentary synthesis of "inner-outer" is a moment where in the same dimension co-presence comes together whereas ordinarily there is a resistance to this. The diagram in Figure 10 illustrates this situation. What is ordinary here is the initial "distance" between the perceptual and the imaginative modes, although also ordinarily focal attention will be weighted upon one or the other. But in the occasion of a synthesis (b), the moment of the particular "hallucination" of a "blending" of co-present awareness, this "distance" disappears.

But in the light of the resistances offered by perceiving and imagining the same thing in the same dimension, a complication must be

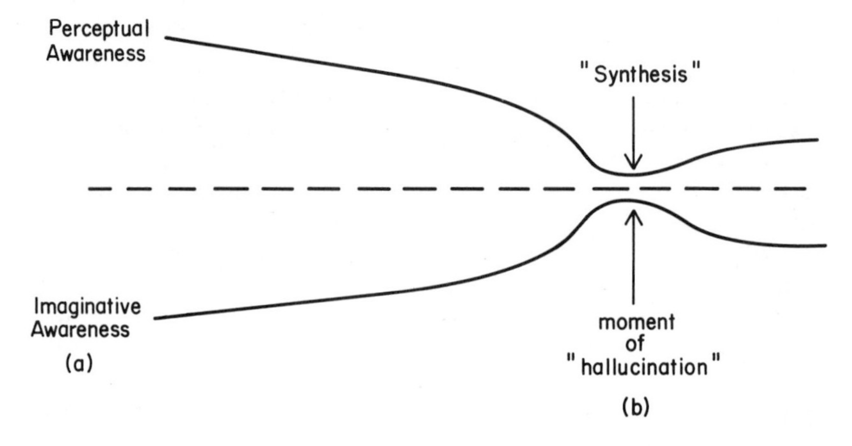

added which in turn serves as a further indirect index for locating the role the auditory plays in "self-awareness." The resistance posed by any synthesis in the same dimension of imaginative-perceptual experience is also a preliminary index for locating at deeper levels the roles of the dimensions of "inner" imaginative activity.

Thus while I am listening to music I may well imagine the flowing of colors, or, conversely, when I am looking at a painting I may auditorily imagine the baroque music which I believe goes with the scene. But when I try to both imagine and perceive the same thing at the same time, *in the same dimension of experience* I immediately run into a resistance which gives way to the alternations previously noted. There is a resistance of the "identity" of the noema which is at stake here, but implied reflectively there is also the problem of the "identity" of the noetic act.

Before taking a final step toward isolating the role of auditory imaginative activity within the full range of the imaginative and perceptual polymorphy, a brief detour into what may have become troublesome regarding the survey of the imaginative mode seems appropriate. The review of the auditory terrain within the imaginative mode almost inevitably raises the question of "introspection" in respect to such investigations. And to compound the problem there is also a question of "contingency." Although the objections associated with such questions often, if not always, betray an already strongly sedimented and often dogmatic "metaphysics" which presupposes distinctions foreign to phenomenology, it must be admitted that there are indeed different empirical habits of thought.

Phenomenology in its own way, however, must always plunge into "contingency," because the existential possibility shows itself only in the midst of "contingency." In terms of this "contingency" it is well established that certain persons have vivid and almost continuous visual imaginations, while others have no visual imagination at all. Others still are focally "linguistic" in their thinking. I think philosophers are often primary populators of this possibility, and more than one colleague has spoken of "thinking propositionally." Moreover, in anticipation of the turn to auditory imagination in relation to language, it must also be admitted that there are "other" types of "languages" than spoken ones. This problem will be addressed below, but by way of preliminary considerations the overwhelming "empirical" observation that most "first" language is a matter of speaking and listening cannot be sheerly ignored. For what lies at the base is again the vast alternative posed by phenomenology *or* metaphysics. For metaphysics the problem of "contingency" more often than not serves as the excuse to overlook the *incarnation* of "thinking" in the modulated forms of embodied experience. For phenomenology the plunge into the midst of "contingency" is a matter of searching out the essential possibilities of how embodiment occurs.

The question in the midst of "contingency" is one which is gradually focused upon the auditory dimension of experience. But at the same time precisely determining a sense of the imaginative is necessary. If there is some sense in which some form of co-presence is continuously or almost continuously present, and if the isolation of a dominating form of that co-presence in its appearance as an imaginative act is to be clarified, what is called for is a more and more "concrete" imaginative variation.

Symptomatically the "concreteness" of the variation may be illustrated in relation to approximating the experiences of those "unlike" ourselves. The situation of the blind person, who has already been noted as having a plenary quality to his experience; yet who is also poignantly "dominated" by an unknown visual dimension which invades the very depth of his social existence, cannot easily be "imagined" by one with sight. As soon as we attempt to genuinely "imagine" ourselves into this state we find a gradual "subtraction" of visual aspects will not do. A closer approximation might be the previously "constructed" imagination which places us suddenly in the midst of creatures who have a "sense" other than ours about which they communicate, but about which we have no idea. Nor is it that we cannot merely "imagine" what they "experience," it is that the very fullness of our experiential plenum excludes this possibility. *There is no way of exceeding horizons except totally.*

In contrast, to begin in the midst of "contingency" is to grant the actuality of that "contingency" but also to open the way to the most "concrete" of variations which show existential possibilities within human experience. The more narrow concern with auditory imagination already preliminarily suggested as a central dimension of human *thinking in language* is the place where that further plunge into "contingency" may continue.

1. Martin Heidegger, *Poetry, Language, Thought*, trans. Albert Hofstadter (New York: Harper and Row, 1971), p. 26.
2. Edmund Husserl, *Ideas: General Introduction to Pure Phenomenology*, trans. W. R. Boyce Gibson (New York: Collier Books, 1962), p. 182.
3. Maurice Merleau-Ponty, *Phenomenology of Perception*, trans. Colin Smith (London: Routledge and Kegan Paul, 1962), p. 183.
4. One graduate student was first only gradually aware that she lacked visual imagination. Later, due to exercises in the use of visual imagery in variations, the student began to doubt whether she could do phenomenology. This may show more about implicit visualism than about phenomenology as such. She later performed brilliantly on auditory variations.

CHAPTER TEN

AUDITORY IMAGINATION

Not all auditory imagination assumes the form of inner speech. There are also the varied possibilities which surround thinking in a language, and which, without investigation, could hopelessly confuse the issue. In the most general terms, auditory imagination as a whole displays the same generic possibilities as the full imaginative mode of experience. Within the active imaginative mode of experience lies the full range from sedimented memories to wildest fancy.

In memory I can recall the voice of my grandmother's quaint Germanic "oncet's" and my grandfather's mumbled dinner grace beginning with, "*Komm'n Sie Jesu.*" In fantasy I can presentify and represent the sounds of the world. I can imaginatively hear the strains of a flute or a cello or both, or I can imaginatively fantasize a debate between two of my colleagues who are not on speaking terms.

Within the realm of the imaginative, auditory imagination may accompany other dimensional presentifications. I recall looking at a National Geographic map of the Middle East, and it presents itself irreally in the imaginative mode. But to it I "add," while recalling the myriad faces of the peoples, the strains of a Near-Eastern wailing melody I once heard.

Both of these presentifications may then be "released," and they disappear. But I seek out the peculiarities of the auditory dimensional characteristics. I notice that there are distances and resistances between the imaginative and perceptual modes of experience regarding co-presence. There is a ratio of focus to fringe in the dual polyphony of perceived and imagined sound. Perceived sound, as in the case of "white sound," or programmed background music, floats lazily around me, and I find I can easily retire into my "thinking self" and allow the floating perceptual presence to recede from focal awareness. But a series of variations illustrates that there are also distances and resistances in the polyphony of perceived and imagined sound.

If suddenly the sounds of the environment increase in intensity and volume, particularly if not constant, I begin to find a resistance to the maintenance of "inner" focus to "outer" sound. The perceived sound in its penetrating capacity disturbs my train of thought. Auditory interruptions of "thinking" are particularly noticeable. A sudden noise, the annual engine trial of someone's hydroplane on the harbor, poses a serious distraction. I recall when living in another town the intrusions of the attempts of a nascent rock drummer whose practice sessions at an open window across my backyard made the truth of the statement *It's so loud I can't* hear *myself think* apparent. The intrusive presence of sound may penetrate into even my "thinking" self-presence.

A second variation of the "disruptive" quality of sound on the occurrences of auditory imagination and the continuities of "thinking" comes more pleasantly in the enchantment of music which can also overwhelm inner self-presence. In its sometimes orgiastic auditory presence the bodily-auditory motion enticed in the midst of music may lead to a temporary sense of the "dissolution" of self-presence. Music takes me "out of myself" in such occurrences.

Each of these variations revolves around the penetration of sound into the very region of the "thinking self." But while sound poses a threat of seduction in some of its occurrences which intrude "inwardly," there are also possibilities of a co-present polyphony of auditory experiences of the perceptual and imaginative modalities. Here the variations begin, however, to leave the realm of strictly ordinary experiences and move toward more extreme variations. However, many of the following experiences are better known, and some are quite familiar to the musician whose auditory imagination is often better tuned than that of the nonmusician.

There is, in auditory imagination, the possibility of a synthesis of imagined and perceived sound as noted previously in a visual example. But in this case the auditory "hallucination" is not a matter of hearing one thing as something else but a matter of a doubled sound, a synthesized *harmonic echo*. I listen to a record of Vivaldi's "Four Seasons." In my new intensified listening I pay particular attention to the trailing off of sounds, following them in Husserlian fashion in their reverberations which meet the horizon of retentiveness. I "hold on" to these notes as deliberately as possible as they trail off. After some attempts at "stretching" attention in this fashion, suddenly and spontaneously there occurs a fully "doubled" passage in the form of a harmonically synthesized gestalt. That is, the notes that were "trailing off" return, doubled as co-present with the next phrase as if suddenly *two* orchestras are playing, one slightly out of time with the other. But the momentary co-presence of a "now-point" with the "just-past" sounds occurs as a full harmonic echo. Later I find that musicians following this period of composition had actually written such effects into their music.

Less dramatically, a variation of the above possibility is more easily detected not as a harmonic synthesis but as a fading reverberational echo of tones just passing off being vividly retained and "added" to tones coming into presence with a definite sense of "distance" such that the echo reverberation is distinguished from the oncoming sounds but also remains as a fringe effect. In exercises of extremely intense listening the doubling effect can produce dissonances as well as harmonies. However, in both the above-noted examples it is unclear what other roles an imaginative modality plays, if any, because the situation described above is close to being an auditory equivalent to the doubled sight which occurs when one crosses his eyes.

In all of the variations upon auditory polyphony cited above, forms of co-presence show themselves as variations upon harmonies or dissonances, upon musical sound. But if I try to imagine and perceive the same sounds at the same time, I find the same *resistance* previously noted. Again the sense of rapid alternations shows itself as the closest approximation to this lack of distance within auditory experience. In this there is an essential isomorphism of the structure of intentionality within perceptual and imaginative listenings.

Further variations begin to show related polyphonies which double perceptual and imaginative possibilities in different ways. I attend a concert, and while it is playing I begin, in fancy, to "embroider" the perceived piece of music with co-present imaginative tonalities. With some practice it soon becomes possible to create quasi synthetic dissonances, adumbrations, variations upon the actual themes being played. There is some evidence that this "distracted" though intense listening may have been practiced by Mozart, who was always accused of never listening to anyone else's music but was busy creating his own version of it even in the presence of another's music. In this form of co-presence there remains a slight sense of distance between the modalities either in the sense of one being the "echo" of the other in a version of foreground and background attentiveness or in the form of alternating bursts of perceptual and imaginative sounds.

In all of the above-noted examples of auditory polyphony the forms of co-presence maintain at least a minimal distance. The perceived sound is in harmony with or in dissonance with the imagined sound. A much stronger resistance is found in trying to perceive and imagine the same sound simultaneously. Here, if any success is achieved at all, close reflection shows a series of rapid alternations between focally perceived and imagined modes of experience. Here a clue is offered regarding an essential isomorphism of intentional structures in the perceptual and imaginative modes.

In spite of what historically has been a massive lack of philosophical attention to the phenomena of auditory imagination, the development of its possibilities, particularly in music, is worthy of investigation. For example, auditory temporal significance may be exceedingly accurate. In the case of Toscanini, tapes of original cuts of symphonies which had been recorded twelve years apart showed that his sense of time was accurate enough for the tempo of one symphony to be within microseconds of the other.

If philosophy has largely ignored the musical ear in both its perceptual and imaginative modes, it *has* attended to the problem of the "linguistic." Although there has been a vast amount of work done on philosophical problems of language, little has been done concerning the examination of concrete forms of thinking as inner speech considered as a type of auditory imagination. In part, this phenomenon *as* a phenomenon of a special type of auditory imaginative activity may have been overlooked because of the long tradition of interpretation which maintains a "metaphysical" and "Cartesian" stance toward thought. This tradition takes for granted that thought is *disembodied.* Thus in spite of discussion of "mental word," the persistence of a dualism of "acoustic tokens" and disembodied "meanings" continues.

But there are phenomenologically locatable reasons for the failure to locate "linguistic thinking" in inner speech as part of the auditory dimension. These reasons lie within the fragility and structure of the phenomenon itself. Inner speech as a form of auditory imagination *hides* itself. Yet in this hidden, fragile, and difficult to locate phenomenon are deeper existential significances for the understanding of *human being as language.*

Thinking in a language, inner speech, though hidden, is also familiar. And as in the case of all familiar phenomena the familiarity itself is a bar to thematizing the phenomenon. Inner speech is an *almost continuous* aspect of self-presence. Within the "contingency" of human language it is focally embodied in thought as an imaginative modality of spoken and heard language. As an *accompaniment* to the rest of experience it is a most "inward" continuity of self-presence and the hidden familiar presence of an experiential polyphony.

The first proximate variations within the auditory dimension displayed the intrusive capacity of sound to disrupt patterns and trains of "thought." And the first indications of distances and resistances begin to foster a more positive suspicion regarding the location and role of inner speech as a special type of auditory imagination. However, further variations are needed to make this phenomenon stand out more clearly.

I return to variations upon musical imaginative presence. When involved in presentifying the "embroidery" of an imaginative musical "addition" to the perceived music, I note that my inner speech ceases. I am "in" the music. I discover here a resistance to simultaneously "thinking in a language" and imaginatively presenting music. Contrarily, when, focused totally upon the multiplicity of imaginative phenomena, I find that I can easily imagine the philosopher's centaur while continuing to "think" in inner speech. Such considerations are not conclusive, although as indirect indexes of the isolation of inner speech as auditory imagination they begin to narrow further the region of location.

Each of the above continues to be a variation in the midst of an often confusing wealth of experiential polymorphy. A reverse set of variations in the form of a detour into a pathology of listening serves to isolate indirectly and inversely the embodiment of language in inner speech. Defects of hearing and, most extremely, deafness symptomatically point to both the "contingency" of what is focally the role of inner speech and to the existential importance of the auditory in the human community. Helen Keller confessed that "the problems of deafness are deeper and more complex, if not more important, than those of blindness. Deafness is a much worse misfortune. For it means the loss of the most vital stimulus—the sound of the voice that brings language, sets thoughts astir, and keeps us in the intellectual company of man."[1]

Language "contingently" focally embodied in sound forms the intersubjective "opening" to the World in terms of the linguistic core of language. Two qualifications concerning this assertion should be preliminarily noted: first, the claim is not to be taken to mean that a loss of the auditory dimension makes "thinking" impossible—this is clearly false—but that the loss of the focal capacities of the auditory dimension displaces the "contingent" focus of thought, although thought continues to be embodied in different ways. Empirically it has long been recognized that the problems of deafness are essentially tied to the problems of language, and that such a relation poses the most serious problem for those afflicted. Secondly, there is at least a weak sense in which, unlike blindness, there is never a case of *total* deafness. The gradations of hearing shade off into a larger sense of one's body in listening. The ears may be focal "organs" of hearing, but one listens with his whole body. The folk music fan "hears" the bass in his belly and through his feet, and the deaf child learns to "hear" music through his hands and fingers. There is, usually, some extremely vestigial hearing in the deaf which can also be partly extended through intensive amplification. But the deaf person continues to "hear" in an essentially different way from the ordinary listener in that what to the ordinary listener remains a fringe effect—sounds felt and experienced in the body—is sometimes the entirety of the deaf person's auditory "focus": he "hears" from *only* the fringe.

Not only does this close off the "contingent" *focal* intersubjective language of humankind, it also effects the way in which he "hears" himself. When I speak I also listen to myself. I feel and take for granted the sounds which I hear returning from my voice. This also gives me a sense of how correctly I may be projecting or enunciating.

But it may be that I fail to notice, until provided with the auditory mirror of a tape recorder, that I do not hear myself as others hear me nor do they hear me as I hear myself. When I speak, if I attend to the entire bodily sense of speaking, I feel my voice resonate throughout at least the upper part of my body. I feel my whole head "sounding" in what I take to be sonoric resonance. This self-resonance which I take for granted does not appear on the tape, and I am initially surprised at the "thinness" and the "higher tone" my voice has on the recording. Physically, of course, not only can these effects be measured indicating the effect of my voice on my skeletal and muscular framework. I hear through bone conduction as well as through the acoustical properties of the air, but the two "media" of self-hearing are essentially separate. There is an essential sense in which *my hearing of myself is distinct from all other forms of hearing*. The same is the case in the presence of my "inner voice" which "thinks" in a language.

1. S. S. Stevens and Fred Warshofsky, *Sound and Hearing* (Time-Life International), 1966, p. 144.

CHAPTER ELEVEN

INNER SPEECH

The familiar but elusive character of inner speech as an imaginative modal counterpart to spoken voice calls for its own establishing variations. Some preliminary qualifications, however, are also appropriate. First, it is clearly not the case that all thinking is "linguistic." There are many important and clearly nonlinguistic aspects to the full range of thought. There is, for example, a kind of visual thinking which is possible particularly in the arts, in design, and in certain kinds of geometrical thought. Nor do I contend here that an auditory "linguistic thinking" is in some way necessary to the learning of language as such, because there may be "languages" in other dimensions of experience. In this sense the auditory form of "linguistic thinking" is "contingent." When language occurs in other dimensions of experience, it remains *embodied* (sensuous) language and is molded according to the dimension in which it occurs. Yet within the "contingency" of inner speech as the normative form of "linguistic thinking" the role of an almost constant self-presence carries important clues concerning the role of thought and its relationship with myself and the World.

Naïve reflections are perfectly familiar with inner speech as the phenomenon of thinking *in* a language. Yet in spite of this easily recognized type of thinking there remains a hiddenness and elusiveness to ongoing inner speech. The first reason for such elusiveness is common to all reflective phenomena which deal with intentional aspects. The very intentional referentiality of experience points away from itself toward that to which the intentional reference points. The very structure of an intentional involvement with the World which also obtains for inner speech hides this form of experience itself. Were this not so, intentionality would "get in the way of" the projects and goals which are fulfilled or frustrated in daily life. Inner speech which accompanies these activities does not intrude itself into them but recedes as a peculiar kind of background phenomenon which provides a continuity and a "sense" to such activities. The very familiarity of thinking in a language conceals its shape.

There is a second and related reason for this elusiveness. As a type of *language* inner speech hides itself for a second time. Language, insofar as it functions to "let be" or allow *otherness* to show itself also does not call attention to itself in ordinary speech. Words do not draw attention to themselves but to the intended things in referring. This extends ordinarily even to the form of embodiment in which the language is found. Thus in speaking, what is ordinarily focal is "what I am talking about" rather than the singing of the speech as a textured auditory appearance. This is not to say that the singing of speech is absent; it is present but as background which does not ordinarily call attention to itself.

The tendency to miss the sonorous quality of speech is related to the tendency to forget backgrounds and to abstractly believe that one can attend to a thing-in-itself. This peculiar and often highly functional background does, however, present itself in dramaturgical forms of speech such as those found in rhetoric, poetry and chanting, and the actor's voice. In such cases even while there continues to be a "showing through" the spoken language, the embodiment of that language in sound is more keenly noticeable.

The third reason for the hiddenness of inner speech lies in its own way of being self-present which is essentially different from other forms of auditory imagination. But its analogue is locatable in voiced speech. Inner speech is to the full range of auditory imaginative noetic acts as voiced speech is to the full range of auditory perceiving noetic acts. A special series of variations which locates this style of self-presence is called for.

I "hear" inner speech differently than I "hear" other forms of auditory imagination. From the previous variations it is quite apparent that an auditory imaginative presentification of an other's voice may be made. But this presentification is distinctly different from that of inner speech. It is an imaginative "listening" to an *other* which I may recall, fantasize, or spontaneously remember. And when such an occurrence is underway, my own inner speech as the almost continuous self-presence of thinking in a language either momentarily recedes or ceases altogether. That is not what thinking in a language shows itself to be.

But when I turn to inner speech itself, although I recognize clearly that it does not appear as "like" the voice of another, I find it hard to grasp directly. I "catch it" from the fringe; it seems to evade objectification. Only with effort and in a sense indirectly do I gradually entice its significance. It does not come to "stand before" me the way in which the other's voice—or even my own voice as quasi objectified—occurs. In these struggles, however, the indirect "glances" of inner speech begin to release certain aspects of the phenomenon. Inner speech is active, ongoing in its elusiveness, and it seems to be "nowhere" or "everywhere" when noted. This "linguistic thinking" does not present itself as coming from "somewhere" but retains its elusive self-presence as either background or accompaniment to the remainder of what I may be engaged in. In this, too, it carries the significance of not being "other," but rather of being *my* thinking.

Here inner speech, though more elusive than voiced speech, retains some isomorphism with spoken voice which also presents itself as coming not from elsewhere. Rather, *my* voice in its self-presence is felt bodily. Furthermore, as an *active* constitution inner speech retains the same sense of "mineness" as voiced speech. A countervariation indirectly points up this significance. Were my inner speech suddenly to become confused and appear to come from elsewhere (as apparently happens in some cases of schizophrenia) I would be startled and confused. In such a case the alienation of inner speech which turns it into an imagined rather than an actively imagining voice bespeaks a deeper division of the self which now no longer "hears" itself properly. Inner speech which is thinking in a language is self-present as "my" thinking self-presence in contrast to other forms of auditory imagination which presentify otherness.

If inner speech is marked by the intimate sense of *my active* thinking, it is also quite concretely a thinking *in* a language. Again, while the subtleties of this phenomenon are elusive, that one thinks in a quite concrete language has been noted often and easily enough. It is particularly notable to those who have "entered" more than one language and recognize that to think in one language as compared to another alters the "style" of thought significantly.

I ask myself, In *what* language am I thinking? And, ordinarily, the answer will be English. Although without the question I may have been only implicitly aware of this, my thinking tends to float back to its mother tongue. This may be noted in the extremities of countervariations which attempt to break this weighted centrality of the mother tongue. I imagine "trying" to think in Russian, a tongue which is opaque to me. Perhaps a few words occur, *nyet*, *da*, but they reveal little. Then perhaps I "cheat" as in the movies, speaking with a guttural accent; but the thinking is then only distorted English. Or perhaps I bring to mind a thinking in "noises," but here there is no thinking in a language at all.

I turn to German, which is readable to me although conversationally difficult, and as soon as I attempt to think *in* the modulations of that language I soon find that the weightedness of the mother tongue is obvious as a type of "inner translating" quickly shows itself as the mediation of thinking in a language. Only when I turn to French do I recall those moments where there is an absence of even inner translation. Dreams, conversations, and lectures are experienced in French and I begin to know what thinking in *another* language is like. Here I think in a language with a markedly different style than that of my ordinary thinking in English. But no sooner do I begin to genuinely think in that other language (after the years of struggle in which only approximations of it are made), than it, like my native English, begins to show itself in a transparency which hides its singing.

In the process of entering a second or third language, however, there is an instructive experience. I have to purposely exert an effort both to listen to and to form the words. In such instances the "sounding" of inner speech under effort contrasts vividly with the case of being in a familiar and thus transparent language.

This very ease of thinking in a language in inner speech hides its phenomenological characteristics. But so, too, do the "speed" and modulations of inner speech. Inner speech does not show itself a word at a time any more than does my voiced speaking. It bursts forth in rapid totalities which present themselves as an uneven "flow." And unless attended to specifically it may be hard to recount just what words have been used at all. One does not attend to words as such but to a larger "singing" of phrases and sentences.

Moreover, these may not show themselves as well-formed sentences at all; inner speech is "colloquial" and "conversational." It "jumps" and "changes key" almost constantly. I rather doubt that even philosophers "think" in the argued jargon which appears in their journals. Stream-of-consciousness writers, attentive to such phenomena, although still reconstructing inner speech, better display this flow and associative "play" of the interior. The style of inner speech is not that of finished writing. It does not have the polished, reflective "time" of words which come to stand upon the page.

Although this speed and irregularity of inner speech is reflectively available, it may also be approximated in a comparison with other language speeds. If I attend a lecture, assuming I am neither a speed writer nor a trained stenographer, I find that in taking notes of what the lecturer says, I write down not a verbatim account but a bare suggestive outline. His actually spoken lecture is far "richer" than the few notes which have "reduced" his saying to a skeleton. Similarly, when I am speaking to an other, my thinking inner speech may be racing, running ahead of my verbal speech such that I always seem to have far more in mind than I am able to voice in such occasions, and this is in part due to the relative *speed* of inner speech. There is no translation here of unworded thought into worded thought, although there may be the transformation of a speeded and running ahead of inner speech into a slower and more deliberate voiced word.

Inner speech, actively constituted, speedy, and colloquial, peculiar in its appearance as "mine," also approaches a near continuity of self-presence in ordinary thinking. But this continuity is a vascillating continuity which oscillates between the *filling* of thought, as in the concentrated thinking upon some problem, and the barely if at all detected "*accompaniment*" of other activities, as in athletic concentration.

Nor is the continuity absolute, although it remains a familiar focus within thinking activity. The events which disturb thinking in a language are preliminarily instructive in this respect. I have already noted the disruptive capacity of sounds to interrupt this continuity. The intrusive power of sound to penetrate even ordinary self-presence also disrupts inner speech. But within the realm of the interior I also find that *auditory* replacements display peculiar resistances to inner speech.

As I "think," I decide to presentify a strain of imagined music, but as I do so I discover that my inner speech momentarily ceases, "turns off," or else it resumes in a series of alternations in the interstices of the imagined sound. I try to think in a language and at the same time imagine the previously imagined melody, and again I come upon the resistance.

Yet when I visually presentify the sailing yacht I desire to myself, I find no resistance to the simultaneous commentary upon that yacht in the form of inner speech. Inner speech as a "voice" may accompany the dimensional multiplicities of imaginative experience, but it meets certain resistances in the auditory dimension.

I push the variations further. In imagining a melody I find I can "insert" rapid bursts of thinking in a language in the alternations and interstices which occur in the rapid inner time which is experienced, but the melody and the "voice" of inner speech war with each other for presence and self-presence. I begin to imagine the presence of others who speak in the cacophony of a cocktail party, and here, sometimes and fleetingly, "my" voice seems to join theirs in a chorus. But careful attention soon shows that when this occurs "my" voice has also undergone a transformation, it has become momentarily objectified from the nowhere-everywhere of inner speech and has become "objectified" *as* a voice of an other. "I" have become a "quasi other."

In such glimpses of the auditory manifestation of inner speech as thinking in a language, its auditory embodiment and its intimacy as self-presence shows itself. Its near continuity as an almost constant co-presence with "outer" experiencing, however subdued as background accompaniment, points to further significations of inner speech. The *familiar* and taken-for-granted character of my inner speech functions to maintain a certain familiarity within the environment. The voice of language domesticates the World.

I begin to explore a new territory, perhaps a beach in a strange land. Suddenly I come upon a creature I have never seen before. I am surprised and momentarily *speechless*, both in the sense of not saying anything and that of uttering a short cry. But also "interiorly" I note a brief interruption in the activity of my "thinking self." As I begin to follow the creature with my eyes, at first puzzled with the question, What is *that*? the momentary strangeness is gradually replaced by movements which begin to relate the creature to similar forms of life which I have seen elsewhere. The commentary of the meaningful, even though not yet successful in this instance, begins to refamiliarize the experience. Speech again begins to pervade in co-presence the ongoing experience in the World.

This disruption of the familiar is recaptured once inner speech resumes its accompaniment to "outer" experience. But at the same time, the thing has been *named*, however superficially or metaphorically: "Ah, that is like a ———." Inner speech as thinking in a language "permits" this continuous and familiar way of moving in the World to return. Inner speech, it may be noted, performs *as* language. It is language which names, which familiarizes, which fits something into a scheme and thereby domesticates it. But *as* language, inner speech is the self-presence of language.

It is here that inner speech, a most "interior" phenomenon, may be understood to be intimately related to the most "universal" of the significations of language as intersubjectivity. I live in the presence and the self-presence of language. Inner speech as a *modal core* of imaginative auditory experience echoes the voice of language in the World. Its polyphonous self-presence is in tune with the sounded presence of the World. This self-presence is, in its core modality, no more quiet than the sound of the World. Its life sounds in word.

This polyphony of inner and outer voices, however, is not always an equal polyphony. Speech flits in its main melodies between the inner and outer voices. At one moment speech and the self are present *in* the explosive expression. In a moment of anger I shout, "You bastard!" and, although I may recover my composure in the next moment, at its occurrence my angry intention was authentically expressed in speech. Or, in a moment of delirious joy the triteness of "It's beautiful!" may be the appropriate expression of speech.

In dissemblance or in double meanings, however, a partial polyphony of speech may be experienced. In the disharmony of deceit what I say hides what I think. I think of how I despise another and feel the sense of satisfaction gained from successfully fooling him. Or in a moment of seduction the ambiguous phrase carries with it the intense desire that it be taken as an invitation to further meetings. Here the doubled voice may be discerned on the fringes of the experience of language.

A fully doubled voice is subject to the alternations of concentrated attention. A "divided attention" shows itself as weighted toward one of its focuses. I walk along, mindlessly humming a ditty, all the while thinking in inner speech. Yet once noted reflectively, the discernment of the rapid alternation of the onset of the humming tones followed by the onset of "thoughts" shows itself. Here the near distance of inner and outer soundings show a difference of a wordless musical humming and worded thought.

I try another variation and speak out loud while at the same time maintaining inner speech. Here the sense of competing resistances reaches an extreme. Bursts of thought now clearly occur in the interstices of the words or disappear altogether as I become engaged in the speaking. Or else the actual speaking itself becomes mindless or repetitive while the thinking inner speech carries the weight of significance. The same alternation may be noted in the thinking ahead which occurs when I formulate an answer while already speaking the answer. There is a resistance to the complete simultaneity of inner and outer speech.

For attention to be divided there must be a distance. Within the realm of the auditory a minimal distance for such a division is that between worded and unworded or musical sound. The mindless ditty, the hum, the mouthed phrase allows a partial polyphony at the fringe. But in the region of the word a massive resistance blocks a full duet of the expressing, speaking self. Even the music of the ditty is a partial resistance.

Much easier is the division of attention which allows one to look about, perhaps scanning the paintings in a gallery, while at the same time he is absorbed in a line of thought. True, one is not letting the pictures show themselves in fullness, but after the tour one can easily describe both what one has seen and what one has thought. The apparent "autonomy" of sight implicates its distance from inner speech. There is a sense in which inner speech "allows" the dimension of sight to stand alone before one. The intrusion of the auditory is, conversely, an index to the central role of the auditory in inner speech.

In this respect auditory imagination "lets be" a visual "world." My inner speech does not strongly intrude upon what is seen, and the "objectivity" of the seen resides partly in this permission granted by the meaningful accompaniment of ongoing experience. Word resides in myself in such a way that language "lets be" a World as a significant phenomenon.

PART FOUR

VOICE

CHAPTER TWELVE

THE CENTER OF LANGUAGE

Listening to the voices of the World, listening to the "inner" sounds of the imaginative mode, spans a wide range of auditory phenomena. Yet all sounds are in a broad sense "voices," the voices of things, of others, of the gods, and of myself. In this broad sense one may speak of the voices of significant sound as the "voices of language." At least this broad sense may be suggestive in contrast to those philosophies and forms of thought which seek to reduce sounds to bare sounds or to mere acoustic tokens of an abstract listening which fails to hear the *otherness* revealed by voice. A phenomenology of sound and voice moves in the opposite direction, toward full significance, toward a listening to the *voiced* character of the sounds of the World.

But there is also a danger of misunderstanding if the idea of language is extended that broadly without qualification. Not only does such an extension risk making everything (and thus nothing) into language, but it places itself in a position of ultimately denying a connection with the philosophies of language and of mind which sometimes secretly share the concerns of a phenomenology of listening and voice. Thus if the extension of the idea of language is taken symptomatically to point up the continuity of all the potentially significant aspects of the voices of the World, then a further distinction must be made which specifically distinguishes the "linguistic" form of language from "language" as the significant. "Linguistic" language is *language-as-word.* It is the *center* but not the entirety of language in the broad sense.

There is some phenomenological precedence for making such a

tactical distinction and characterization of language-as-signification and language-as-word. From the outset the "unit" of meaning in phenomenology is experiential rather than "merely" linguistic. To speak, to understand, and to perceive are *meaning-acts*. But also what is heard, understood, and perceived—within the realm of human being—is taken as already pregnant with meaning. Furthermore, the intentional structure of human-World correlations already shows that there is a kind of functional isomorphism involved in all meaning-acts, and thus, as Ricoeur says, meaning or significance is already *both* perception *and* word.

Admittedly, this tactical distinction blurs the difference between perception and language in favor of meaning-acts, but this blurring is also already latent in phenomenology from the outset. Moreover, this blurring is also addressed to phenomenology in its later forms. To elevate the importance of language is not meant to demean perception which has been the first and often dominant theme of recent phenomenology, but it is meant to purposely move away from the still recalcitrant vestigial "empiricism" within phenomenology. This vestigial empiricism is a matter of a certain interpretation of the levels of experience and of a seeking for origins, often in mythical pasts. There is here a rejection of lost and forgotten pasts, of "prelinguistic" levels of meaning in favor of a beginning which is better characterized as *starting in the midst* of that which is already here, the already constituted. That is a "beginning" from the center.

The notion of a center, however, calls for a preliminary and general location. Center partly, but only partly, relates to the previously developed notion of a *focal core* within some dimension or the totality of global experience. The center of language as language-as-word may be understood as similar to the appearance of a focus-fringe phenomenon in the sense that deployed around language-as-word is a vast field of meaningful activities which may in the broader sense of language be called "languages." These "languages" are gravitationally weighted toward the central significance in word, but they may be relatively distinguished from the "linguistic" form of language.

There is thus a "language" of gesture which can itself be rich and highly significant in its expressibility. Without spoken word Marcel Marceau can mime into existence an entire context which is "silently" understood. The other's *face*, particularly, "speaks" a silent "language" in the smile, the frown, the slight tinge of sadness, or the massive blankness of mourning. This is language-as-significant, but without word. There is also a "language" of actual touch which belongs to gesture. In intimacy the "language" of touch conveys an often greater intensity of intersubjective communication than a word may seem to convey. But after the intimate, beyond the gesture, language-as-word returns as the weighted center of significance, and the daily traffic of voiced speech and listening resumes its functional centrality.

As a center, language-as-word can also be *de-centered*. Thus it is not to be understood as a fixed focal core. The "languages" of gesture and touch may in fact become the focal cores within a duration of ongoing experience, placing language-as-word momentarily to the side. Yet there is the "inner" reassertion which insists on resuming its role even in the presence of an "exterior" silence as in the case of my thinking in a language which may accompany a revery. There is sometimes a hiddenness to the center.

Insofar as language-as-word is the center of language, it never stands alone within the range of the significant. Again, like the notion of focal-fringe structures, the center of language presents itself only in the midst of a wider significance. Word does not stand alone but is present in a field of deployed meaning in which it is situated. In this sense there are always other significations along with word. This is the *co-presence* of word and wider signification. Here other possibilities emerge, possibilities of the overwhelming complexity and richness of the broad sense of language which threaten to subdue the search for structures and invariants and which point to the essentially *open* horizon of language. What is said always carries with it what is present as unsaid. In the co-presence of language-as-word centered within the field of language as the significant there is a range of variations which indicate that "too much" is being said.

At one extreme there may indeed be a harmony in the saying which brings the unsaid significance into a united meaning-act. The child's laughing voice reverberates harmoniously with the look of her smiling face when she receives a gift. But at another extreme there are variations between the said and the unsaid which equally hold the possibilities of dissimulation. He smiles as he speaks, but his unkindness shows darkly through his words in the touch of sarcasm revealed. Here, only he who listens well can detect these subtleties which do not always float on the surface of the words. And he who does not or cannot listen deeply may hear only the words. Further still lies the dissimulation which allows what is spoken to be given the lie by what is thought in a disharmonious co-presence of "inner" and "outer" speech.

A complete topography of language would thus have to deal not only with the lateral relations of its center in the saying to the gesticulatory and contextually surrounding field of significance but also with the depth of relations between the "outward" center of language-as-word and the possibilities of the range of harmony to disharmony among the polyphonies of the human voice. Within this complexity of the relations of language-as-word lie the essential ambiguities of actual speech and life. Thus, too, the experience of language when considered in contexts wider than technically restricted "languages" reveals itself as containing a degree of existential uncertainty and revocability within the ordinary speech of humankind. To listen and to understand mean more than the comprehension of words, they signify entry into a wider communication situation.

However, the dominant problem here is not the examination of an entire philosophy of language nor even the possibilities of existential language. Rather, the dominant problem lies in the auditory dimension, *voice*. Language comes alive in word, but within the "contingency" of human word there also may be seen a functional centrality to *voiced word. Language-as-word is normatively embodied in sound and voice.* If the center of language is language-as-word, that center shows itself in the ongoing traffic of human interchange "first" and dominantly in the auditory dimension. Its significance is a *meaning-in-sound.*

Here, in fact, we meet two problems. First, if language-as-word is the focal center of the languages of the world and the self, those languages are phenomenologically understood to be existentially *embodied* languages. This is to say, with Merleau-Ponty, that the word has a meaning, or, better, the word *is* a meaning.[1] In either case it is the actual, concrete word, the sounded word which is the meaning. In this respect phenomenology remains thoroughly "anti-Cartesian." Its "linguistics" must also be "anti-Cartesian linguistics."

Meaning *in* sound embodies language. But it is not the only embodiment of language. Nor is it even the only embodiment of language-as-word. This is because there is also a possible second de-centering of language-as-word in terms of different embodiments of language. Historically, of course, the most important form of second decentering has been that of spoken word by written word. And although there are minor gesticulatory languages which also decenter the auditory word (as in the case of codified and conventionalized sign languages), the appearance of writing remains the primary "second" embodiment of language-as-word. This possibility constitutes the second problem in the understanding of the role and the importance of voiced language as the normative center of language.

Moreover, these two problems cross in resisting the potential overthrow of "Cartesian linguistics" for the sake of an emergent comprehension of phenomenologically embodied language or existential language. The phenomenology of language finds its justification in the *absence* of fulfillable non-embodied meanings. Where meanings are found, they are found already embodied, although the variations upon embodiment are complex. The word is sounded, seen, felt; and even in thought its presence takes its own "shape," whether in inner speech or in the soundless presences of other dimensions of the imaginative process.

But there are two preliminary directions in distinguishing embodied language from its "Cartesian" lineage. The first direction is that of a movement toward the fulfillable experience *of* language which is simultaneously a movement away from hypothesized "disembodied" meanings. In the auditory dimension this is the movement which reflects upon the presence of listening.

The first direction is that of a temptation to accept as heard the posited "abstractions" of Cartesian metaphysics. When I listen to an other I hear him speaking. It is not a series of phonemes or morphemes which I hear, because to "hear" these I must break up his speech, I must listen "away" from what he is saying. My experiential listening stands in the near distance of language which is at one and the same time *the other speaking* in his voice. *I hear what he is saying*, and in this listening we are both presented with the penetrating presence of voiced language which is "between" and "in" both of us.

A "Cartesian linguistics" however, does not hear. It supposes its listening to hear "bare sounds," "acoustic tokens," which in an undiscovered "translation" are mysteriously or arbitrarily united with the disembodied and elusive "meanings." These meanings float above and beyond the embodiment which is what presents itself to listening. Experientially just as the thing is always already found as the "unity" of its "qualities," so in language is the word always found already embodied and significant.

But to listen with a phenomenological "naïveté" is by no means simple. The infection of a "dualism" of the "body" of language in abstract sounds with its presumed disembodied "soul" of meanings pervades our very understanding of listening. This linguistic dualism constantly tempts the listener to hear what is not heard. Thus to more fully locate the fullness of voiced word there is a need to take note of the near and far reaches of sounded significance which remain "outside" language-as-word. On the near side there remains the enigma of musical presence, a sounded significance which is nonlinguistic. On the far side there remains the enigma of the horizon of silence which situates, surrounds, and permeates the presence of word.

If the first temptation of a "Cartesian linguistics" is the dualism which leads to an abstract listening which is no listening, the second temptation is to see in other embodiments of language-as-word an alternative which relativizes the spoken word and relegates it to a mere "contingency." Ordinarily, in literate cultures, humans are adept at two embodiments of language-as-word, the spoken and the written word. But just as there are differences of arrangement within the dimensions of perception and imagination, there are also differences in the double embodiments of language-as-word. The *shapes* of the focal-field arrangements in voiced and written language are not perfectly isomorphic in their characteristics. There is a difference in the mode of presence regarding significance.

If I write the word *Adam* alone on a page its context remains opaque. For the reader who comes upon the word on a page, the field and its unsaid significance is a dark obscurity. Perhaps, if I am a philosopher, I surmise that this is a "bare name." But if this word is *spoken*, there is already a certain potential field and presence of unsaid significance in the voice. If "Adam" is said in an angry voice, imploringly, or in a quiet whisper, each sounded presence allows the "bare word" to emerge from some of its obscurity in the sounding of its presence.

This is not to say that the "same" context and presence of the unsaid could not be elicited in written language-as-word. But in that case there is also a significant difference in the mode of presence in which the context and the unsaid occurs. I can, as I have done above, fill in the opaqueness by adding more words. I can surround *Adam* with the context I wish if I am skillful in my writing. I can write "Adam" to be said with staged direction, imploringly and thus suggest the sounded word context which I hear immediately in the spoken word.

However, in both the versions of a word in context there has been an embodiment in sound or in that which is sighted. Without embodiment the "meaning" does not occur, but with embodiment there is a difference in the "sameness" of meaning as a phenomenon. It is here that the usual meaning of "reducing" speech to writing occurs. The written word "lacks" the sounded significance which already gives a degree of context to the word, and this makes the unsaid less opaque. Writing fails to convey that minimal sense, although the "reduction" can be compensated for by adding words which also soundlessly replace, in their own way, what was lost. Writing creates the possibility of a *word without voice*. It opens the way to the forms of unvoiced word which secretly dominate whole areas of the understanding of language. Husserl foresaw this relationship in his article, "Origin of Geometry," insofar as he discerned that the higher reaches of mathematizing thought (as more advanced forms of "voiceless" language) are in fact dependent upon the emergence of writing.

Humans have not always recognized the possibilities of voiceless language. The fourth-century student who came upon Ambrose reading in the scriptorium *without saying the words* was amazed at "silent" reading. Nor does everyone, even today, read without the noticeable presence of outwardly silent but inwardly "sounded" words. There is here a continuum between voice, voiced reading, inner voice only with reading, and voiceless reading. Yet even voiceless reading can subtly reestablish its secret liaison with the adherence of the spoken word when the phenomenon of "hearing" a friend in the book which he has written occurs.

The secret adherence of speaking to writing remains in the learning of reading as well. The realm of written language-as-word is entered as a "second language" after a person has already entered the realm of speaking. At first the words are "sounded out," and the reader learns a written language as slowly and as simply as he did first speech. But once having entered the second embodiment, word without voice becomes possible. This possibility is also one which moves further and further away from the liaison with voiced word.

At its most suggestive and descriptive, as in the novel or the poem, writing still reflects and elicits a sense of the auditory. The characters of the novel breathe and feel and speak, and the imagery of writing allows the sensory adherence of life to show itself. But in the language of the report, of the newspaper, the account, no voice emerges; or if there is voice it is personless as the "everyman" of Heidegger's *das Mann*. Without voice the *per-sona* recedes, and there is the possibility of "depersonalization."

The unvoiced word of written language, however, is but a first existential possibility of decentered word. A second movement is also possible in the heart of unvoiced word, and that is the step toward both *unvoiced* and *wordless* "language." Further from speech lies the realm of the voiceless and wordless "languages" of logic and mathematics. The first decentering of language-as-word which eliminates the voice of the other opens the way to a second possibility which eliminates the word itself. But the word may be eliminated only by embodying "language" differently, in the abstract symbol or the number which now replaces word but remains embodied in its own mode of presence.

There is born in the graded possibilities opened by the "second language" of writing a progression toward "languages" which are *distant* from language-as-word. But this distance is deceptive, because the movement away from the center also retains a relation back to the center. For even as the new pathway of thinking opens there remains a weighted center of gravity from which the pathway moves. The thought which takes its flight from voice and word remains situated in relation to the thinking-speaking which is "ordinary" language.

Not only do the logician and mathematician return, for their ordinary affairs, to the region of spoken word, but the "meta-language" which situates and surrounds the flight from voiced word itself remains the forgotten, implicit context from which the flight takes place. When I, thinking as a linguist, "objectify" language and posit it as a system of sentential structures with "grammars" whatever their surface or depth; or when I, thinking as a logician, follow the deductive movements of a wordless calculation, I also find already present and embodied that thinking in my mother tongue which floats among my other thoughts. My "objectifications" relate back to and presuppose a living language within which I already stand.

If this is the movement which is opened by the decentering of voiced word on behalf of written word, there is also the countermovement in the opposite direction. Just as there are voiceless words, there are wordless voices, the voices of things which are a wordless speaking. Such voices are pregnant with significance but not yet word. Thus I recognize the voice of the truck which sounds differently than the car, or the voice of the neighbor's dog which barks differently than the occasional harbor seal which also "barks." The voice of each thing bespeaks something of its *per-sona.*

Here lie possibilities of another extension from the center of language, perhaps not so fully developed as the first flight away from voiced word. There are anticipations of these possibilities in the ordinary "understandings" which occur, for example, between humans and animals. My dog "understands" something from my speech: she knows her name, even when it is spelled (which she has learned to recognize). And in the new sciences there is hope of "understanding" dolphin or whale "language" sometime in the future.

But in this direction often lie only the projections of our worded voices extending to speechless things a certain hearing of speech where there is no word. If awaiting a guest in the summer, the wind and more particularly the "babbling" brook carry sounded voices for my keen anticipations. Try as I may, I do not succeed in eliminating these occasional "voices" in the "babble."

In this direction lies the seduction of the *musical* which is near language-as-word. Music embodies significance in sound, but it is the sounded counterpart to the wordless "languages" which arise out of the possibilities of written language-as-word.

In the history of phenomenology it has been Merleau-Ponty above all who has pointed to the intimacy of language and music. To speak is first to "sing the world," he affirmed. And in the nearness of music to language, the incarnation of "meaning" *in* sound seemed most clear in the case of music where there can be little doubt that the meaning *is* the sound. Merleau-Ponty found even the apparent opacity of music to be closer to language than it is usually thought to be, for the "grammar" of a musical piece fails to yield the "transparency" of language only in its self-containedness. The sediments of the conventional do not so easily transform themselves in music as such.

But some music while "close" to voiced language and "closer" than other forms of the wordless, remains wordless. The dark mystery of music shows itself differently, and the listening which it calls upon is not the center of voiced language-as-word. Music in its nearness to the center helps locate the center.

CHAPTER THIRTEEN

MUSIC AND WORD

In all music, sound draws attention to itself. This is particularly the case in wordless music, music which is not sung. Here the "meaning" does not lurk elsewhere, but it is *in* the sounding of the music. There is even a sense in which the listening which music calls for is a different listening than that called for by word. Wordless music, in its sonorous incarnation, when compared to language is "opaque," as nothing is shown through the music. The music presents itself; it is a dense embodied presence.

It is this *immediacy* of music that Kierkegaard described as the "pure sensuous," the "demonic." But there is no "purity" to the sensuous, it is rather a matter of the pregnancy of meaning which presents itself as music. Only when compared to language, the center presupposed long before being called a center, does music appear as a "dark mystery." Nor, for the same reason, can music be thought of as "abstract"; there is more than a surface to the sound.

Phenomenologically the question is one of listening reflectively to what occurs, to how music is presented if the pregnant significance is to be detected. If music in its unworded form does not "refer" to the world, if it is not characterized as a "transparency," its mode of presence must be located otherwise. But in this, music is not different from other sound presences, although it accentuates and emphasizes possibilities in its own unique way. Its "reference" is not things, but it enlivens one's own body. To listen is to be dramatically engaged in a bodily listening which "participates" in the movement of the music. It is from this possibility that the "demonic" qualities of music arise.

In concentrated listening its enchantment plays upon the full range of self-presence and calls upon one to *dance*.

Dance, however, must be understood not merely in a literal fashion, for dance in this context is the enticement to bodily listening. Thus the full range of the dance to which music issues a call is one which spans the continuum from actual dancing, as in dance music or in the spontaneous dances found in rock festivals or religious revivals, to the "internal" dance of rhythms and movements felt bodily while quietly listening to baroque music.

It is in the call to dance that a different reason for the Cartesian temptation to conceive of sound as a "body" emerges. If, on the one hand, music is sound calling attention to itself, the temptation then is to conceive of music as "pure body"; but on the other hand the call to dance *does* engage *my* body But what occurs in this engagement is clearly anti-Cartesian. It is my subject-body, my experiencing body, which is engaged, and no longer is it a case of a deistic distance of "mind" to "body." The call to dance is such that *involvement* and *participation* become the mode of being-in the musical situation. The "darkness" of music is in the *loss of distance* which occurs in dramatically sounded musical presence.

Not only is one's bodily sense engaged, but the previously noted *filling* of auditory space occurs as well. The now dramatic sounding encompasses and penetrates listening. This filling of auditory space is the loss of distance, of an open space to listening. It is a form of musical ecstasy which is at the other end of the possibilities of "objectification." Music *amplifies* a participative sense of bodily involvement in its call to the dance.

But such dramaturgy of musical sound is not absent from any other experienced listening, although it may be withdrawn and minimal. If I return to the realm of word, I can detect similar losses of distance in the filling of significant auditory space. His angry shout can electrify me, and I feel the threat in his tone. Her whisper is an enticement which sounds irresistible. In sound presence there is always this possibility. In spoken word there is a *dramaturgy of voice* which is essentially musical. Music amplifies the dramaturgy of sound.

If music is thus used to locate an aspect of voice, in the realm of word this musical presence may be heard, not as absent, but in a different way. Ordinarily we think of speech as primarily a matter of communication, of a "transparency" toward something that is not itself speech. This is not always the case. There is, in the learning of language, a movement from the dominantly musical to the later "transparency" of a language already learned. The Vietnamese language sounds tinkling and bell-like even if what is being said is a curse. I recognize in German a singing which, before I enter the "transparency" which only gradually occurs, is a musical phrasing which is more like that of English than that of the French tongue. The foreign tongue is first a kind of music before it becomes a language; it is first pregnant with meaning before the meaning is delivered to me.

Inversely, there is an analogous sense in which music also has a "grammar" and a style. No one mistakes the Rolling Stones for Mozart, nor do the more learned mistake Handel for Haydn. Yet all of these musical "grammars" are closer together than the strange "grammar" of gliding, complex, and stylized pieces of Indian music which to the beginner first often appear as not even "music." The sounding of sympathetic strings and the use of twenty-two microtones, the whine of sitar and sarangi, present musical confusion. Yet once learned, Indian music proves to be one of the most highly classified, organized, and hardened musical traditions in musical history.

These approximations, however, do not yet precisely describe the nearness of music and word. The "music" of language and the "grammar" of music remain caught in a metaphysical classification. There is a sense in which, phenomenologically, spoken language is at least as "musical" as it is "logical," and if we have separated sound from meaning, then two distinct directions of inquiry are opened and opposed. But in voiced word music and logic are incarnate. No "pure" music nor "pure" meaning may be found.

Yet, except in clearly dramaturgical situations, the sounding of word does not call attention to the sounding as music does. In ordinary speech the sounding of word remains in the background. This is not unimportant. The vibrantly expressive speaker is usually thought to be more interesting than the dull speaker. The difference is sounded. The strong voice commands where the thin and wispy voice does not. Yet the sounding withdraws as the context and setting in which *what is said* emerges as foreground.

Here the "darkness" of the musical yields to a "transparency" of a particular type. It is the "transparency" which is located in the *enabling power* of word. Sound in word "lets be" what is not sounded. A return to an artificial approximation in a "Wittgensteinian" example may begin the establishing variations.

Suppose first in a "Wittgensteinian" language situation that meanings must appear as words, but that these meaning-appearances also must be visual. In such a case, then, as I walk down the street I might come across a sycamore tree, but it would have no meaning for me unless its appearance occurred with a word. Thus as I look at the tree suddenly a translucent "word" would appear in front of the tree, and the tree would be properly "named." But in this case the translucent "word" would also intrude itself between my seeing the tree and my recognizing the tree as tree.

But now suppose that the embodiment of the word-meaning could occur in sound. This time when the sycamore tree appears I look at it and hear the name *sycamore tree*. In this instance the tree may continue to stand before me undisturbed while it receives its meaning in the sounded word. The sound-meaning does not disturb its visual presence but lets its visual presence be.

The example is overly simple. In ordinary experience it is clear that every time I see something I recognize I do *not* say to myself, "Ah, that is a sycamore tree." Although I could do this, and in cases of doubt or ambiguity I may indeed begin an inner monologue, what the example does begin to indicate is that the possibility of a near distance is opened between an embodiment of meaning and that to which it "refers." If the "referent" is in the same experiential dimension as the embodied meaning, one might be led to expect the resistances of doubled sameness which I noted previously.

An objection, still in the context of the overly simplified "Wittgensteinian" example, can also occur. If the role of sounded language is to "let be" the visual appearance, then would not one also expect meaning embodied in sound to pose a similar problem to listening? The answer is, of course, that such is indeed the case, but it is the case in terms of polyphony. In genuine listening to another in a conversation I must let him speak, I must resist both speaking and allowing my own inner speech to intrude. Within the polyphony of the spoken and inner speech, if I begin to think along my own line of interest while he is speaking, I find that his speech recedes and that I have to reconstruct it from the fringes of the auditory dimension: "What was that you were saying?"

Between the visual appearance of the tree in the approximation and the sounded inner speech there is a *difference* which may be called a near distance. In this distance there is a clue for the lack of attention which often occurs in the philosophy of language to such phenomena.

The role of sound does not point to but away from itself, "allowing" what is *seen* to stand out. Language-as-word, unlike music, even while sounding, does not draw attention to itself *as* sound. And yet, were the other to be speaking, and suddenly the sound actually disappear, I should no longer be able to hear what was being said. The "transparency" of his speaking would not merely be diminished but disappear as explicit. What was being said in the sound retreats and becomes opaque, but significance does not disappear entirely. It is transformed to the vague and implicit significance which I can *see*. In watching the silent spectacle of his speech I see that "something" is being said.

Here a second oversimplification in the example emerges. In its present form it implicitly denies significance to the visual elements of the approximation. When I meet a friend I do not spontaneously say to myself, "Aha, there's Bob," although I may do so if I wish. I do not need to "say" this to myself, because Bob's visual presence is already significant though not expressed nor situated in a line of thought. When, however, he enters the conversation, then the alternation of polyphony and of central signification as word returns to its normative position.

The possible dialogue which can then commence displays a series of complex possibilities regarding the modalities of sounding presence. Ideally, perhaps, he speaks, and I am silent both vocally and inwardly as I listen to his voice. Then the opposite occurs when I speak. I am "in" my voice, and there is no echo from the inner polyphony which I can master. But there is also the possibility of only partly hearing what he says as in the case of an intrusive inner speech, and the same possibility applies to his listening. But in each case the degree of intrusion is in terms of the added sound presences which occur in the communication situation in relation to language-as-word. In the dimension of sound the situation of an intrusion of the same order does occur. The enabling power of word occurs in the midst of the fragility of polyphonous sound.

Music locates the function and role of the center of language in a second way. Music draws attention to sound, but its sound is a transformed "created" sound. In this sense its sound is *strange*. Its sounds are not those of unattended things nor those of spoken voice, although music shares intentionality with human voices. To an extent this strangeness of music applies even to the contemporary experiments which seek to escape "composed" or "constructed" sounds. The music which is a mélange of "natural" sounds draws attention to the musical character of all sound—(I have already noted this aspect). The aim is a transformation of listening, a listening to the music of the World.

Ultimately, however, such a listening does away with the idea of music as such, for then music is not distinct from sounding in any appearance. As "set apart" however, music retains a certain strangeness. Each new piece, each melodic gestalt, provides a "new language." We do not listen to music all day, its time is "set aside" as a special time.

In contrast, the presence of word is *familiar*, and its sounding is that of a familiarizing continuity, particularly in the ongoing self-presence of inner speech.

But there are times when this familiarizing continuity is broken. A most dramatic occurence is shock. A phenomenology of shock would show that during the moment that shock is incurred there is a suspension and disruption of the familiar in an extreme way. What ordinarily appears as stable, understandable, and structured at the moment of shock becomes disorganized, chaotic, fluid, and lacking in the hierarchies of value and meaning experienced as the ordinary. Shock occurs as an *absence* of familiar word, and talk may be needed afterwards as a therapy of recovery.

Less extremely, the strangeness which may occur in the *absence* of word may be shown in deliberately constructed experiments. I ask my students to do variations upon the exercise, "Try not to think of a white bear." Each case is a deliberate attempt to "turn off" inner speech as linguistic thinking. This results in several outstanding occurrences. First, there is a certain difficulty in deliberately "turning off" inner speech which indirectly demonstrates its familiar but nonobtrusive self-presence. But with effort and imagination, ways may be found to accomplish the "turn-off." Some of these have been noted as the interruptions to the flow of inner speech.

A second step is then called for and the experimenter is asked to attend carefully to the momentary appearances of things in the interstice when inner speech is silenced. Again the response is quite uniform: in the moments when this occurs things become unfamiliar and strange. The descriptions which result often include terms such as "more vibrant," "alive," "unstable," and even "uncanny." These descriptions, when the presence is one of animals or of persons, sometimes mention a sense of "power" or a feeling of "fear." Familiarity is displaced, and strangeness is found to be lurking within the very nearness of things in the absence of word.

These descriptions, however, are not unknown to philosophy, even if they are extraordinary for descriptions of things, animals, or persons. They are, rather, closer to the classic descriptions of intense aesthetic, religious, and mystical experiences recorded in the history of thought. These are usually thought of as exceptional experiences, but in the cases mentioned above what was exceptional was enticed by a purposeful thought-experiment.

Phenomenologically, it is more appropriate to term such experiences *horizonal.* They are experiences which "stretch" ordinary experience to limits. The wordless presence of strangeness indirectly shows the more ordinary function of familiarization which the presence and self-presence of word allows. Inversely, the revelation of strangeness lurking within the presence of things which is ordinarily concealed by familiarity points to the need to develop further the relation of horizonal phenomena to the center of language.

The absence of word in momentary occurrences elicits the significance of the horizon as *silence*. Silence is in some sense a limit of language-as-word, a limit which constantly withdraws from the center. Word is present, but as situated within a wider field of signification it reaches outward to the ultimately silent horizon. Within the ordinary the horizon hides itself, but at the limits the horizon has its own way of revealing itself.

CHAPTER FOURTEEN

SILENCE AND WORD

The horizon as silence situates and surrounds the center. This is the meaning of horizon as first outlined in the approximations of the auditory dimension. In this respect the horizon at its extremity first shows itself (indirectly and at the extreme fringe) as *limit* which trails off into the *nothingness* of absence. As extreme limit the horizon constantly withdraws and hides itself, yet it is that which situates the entirety of presence itself. Horizon as limit and horizon as the Open is thus the extreme degree of possible description.

There is, however, a third significance for horizonal phenomena which is closer but more hidden, which must be drawn upon for its role in further locating word as center. This is the horizon as the unsaid, the latently present; horizon in the midst of presence as the hidden depth of presence. To return to modeled approximations which elicit this sense of the horizon, a return to perception may be made. Things show themselves as "faces" but never as mere "faces." They are situated and hide within themselves as latently significant another side. This is a significance which I implicitly recognize and expect: I am not surprised when the block is turned around and it shows a different "face." The thing presents itself as having a back, as having a depth. This may be spoken of as a local or latently present horizonal feature of the thing. It is the hidden side of presence which is enigmatically "in" presence.

Again the approximation has been primarily a visual one, so the next step is to locate the same feature auditorily and, in the present context, in terms of word. The voiced word, however, also shows itself as having a hidden depth, a latently meant aspect. This is concealed within but detectable in listening to language. In everything *said* there is the latent horizon of the *unsaid* which situates the said. Yet, as in all horizonal phenomena, the horizon is that which withdraws. It is easily overlooked or forgotten. Easy or naïve listening attends only to the center, but in doing so the latent meaning of the horizon remains taken for granted and its latent meaning situates the saying by its unsaying.

The variations which begin to elicit the significance of the unsaid cover a series of horizonal phenomena. The broadest horizonal feature regarding the unsaid as latent significance is the feature of the *unspoken context* which surrounds speech. The context belongs to a degree of silence. Here the variations which most pointedly mark the horizonal role may begin in situations of opaque contexts. If I begin to speak to the other in terms of *halyards*, *sheets*, *gybing*, or *bending on a line*, the listener who has not yet heard the "language" of sailing may return a blank, puzzled stare. I have said something to him, but he has not heard in my saying all that is to be heard. Similarly, in the midst of the tribe of philosophers, if I begin to make these notes on the board, $p \supset q$ or $p \vee q$, the instant recognition by the initiated of the wordless "words" of symbolic logic may appear to be perfectly transparent, but to the uninitiated they would be perfectly opaque. In each case there is a border of the unsaid which, until entered, hides the saying itself. In these cases the language also hides in implicitness but is silently heard or not heard in the saying.

The silence of the context, however, is not a blank nor total silence, it is the near silence of what *can be said*. In this the example is similar to the visual example of the latent "face" of the thing. I can turn the thing around and view its other "faces" and see only a relative degree of hiddenness at any one time. There is always some "face" or other which is hidden—the *ratio* is an invariant structure—but I can get any "face" I wish. The same is the case with the low horizonal degree of a near context. This degree of the unsaid may be obtained and heard.

But it is also important to note *how* such a degree of the unsaid may be heard. Its silence is one which implies that in some sense what was not said explicitly *has already been said*. While not all can be said in a saying (there remains a *ratio* to the unsaid which is the transcendence of the context) what was not said has been said in a *community* with a *history*. Existentially implied in the context is some kind of tribe, or community with a history. Learning to hear the unsaid gains entry into this community and history to some degree. The learned is the initiate who has already heard and thus has entered into the community and the history.

There are technical "tribal languages" whose sayings hover near ordinary speech, but in which there are highly determined meanings which are heard only by the initiate and not by the ordinary listener. The unsaid can be missed in unlearned listening.

I wander through the mazes of the university seeking those technical "languages" which deal with auditory experience. I chance upon a lecture in acoustical physics. I listen. The lecturer speaks in English, and the words he utters seem familiar. He speaks of *acoustical reflection*, of *plane reflection*, of *parabolic reflection*, of *elliptical reflection*. Yet although the words are ordinary, their significance does not appear as immediately obvious to the stranger. Lurking at their fringes lie the yet unknown regions of the unsaid, the silence of the presuppositions, and the framework of definition which gathers the saying. There is a certain strangeness to the words. But once the massive unsaid is heard, and one returns to the saying, its obscurity vanishes, and there is a clear, light, and present meaning to the terms. To know a sentence entails knowing a language. This also implicates the community which speaks the "language." To enter the language is to enter a form of life.

The learner must undergo a catechism of definitions and relations in the technical "language." He gradually learns to speak like a member of the "tribe," and in the process the significance of the word becomes intuitive, for he has learned to hear the echoing and reverberating horizonal significance of the unsaid.

The communities and histories which carry variations of the unsaid are multiple and complex. There are "languages" which are also distant to ordinary speech. I enter a church where there is a prayer service. I listen, and the ritual is seemingly in English, but its tone is archaic. I hear spoken *Thee* and *Thou* and perhaps even a reference to "thynges that go bumpe in the nyghte." I am mystified, and the significance which lurks in the ancient words escapes me. But if I become an initiate the unsaid is gradually unfolded. I begin to hear a reverberation from ancient times and from the silence of the past there begins to spring a certain life. Adam, Abraham, and Amos begin to live in pregnant significance. I listen again to the ritual and begin to discern the regioning horizon as no longer opaque but as the echo of the past into the present. For the ritual tongue ties humankind to that which has gone before him. Even, indeed, if the ritual has transformed itself into the "timeless" distance of that which occurred *in ille tempore* as in the ahistorical forms of religions. The days of the gods are to be repeated and remembered, and the ritual spans this distance in its dramaturgy.

In both cases, that of a technical language linked to a scientific community and history and that of a ritual language linked to a religious community and history, there is displayed a ratio of the said to the unsaid. And for both there is a moment in which forgetting this ratio becomes possible. Word does not stand alone but stands in a ratio to the unsaid, the immanent horizon which proximally situates the saying. But the initiated listener can so take what has been said for granted that the clarity and obviousness of what is now said tempts him to forget that this clarity and obviousness has been attained by longer listening. And in his temptation the "truth" of his tribal language is thought to be "timeless." The surface hides a depth.

Not all depths are, however, ratios to the traditions of tribal languages as such. Within the complexities of speech lie also polyphonous significances which are possibilities of the ratio of the said to the unsaid. I am a lover courting my beloved, but we are still partly strangers. What I say on the surface is ambiguous; it is an invitation to share a more intimate liaison, but it is masked in such a way that should she refuse me I may retain my composure. Or I am a politician, and the surface of my words conceals more than it reveals. But the careful listener who knows this language of purposeful ambiguity detects in the slight change in wording the sign of a change in position. Here one listens "beneath" the words, his intention is to hear "below" the surface, and there, guarded but understandable, is the language of the unsaid.

The varieties and complexity of the ratio of the said to the unsaid are indefinitely large in number, and a comprehensive hermeneutics of language would have to address these varieties. For my purpose here, however, it is sufficient to note the nearness of significant silence as a proximal horizonal feature. The listener hears more than surface in listening to word. The clarity or opacity which he discerns in the saying remains in part dependent upon the learning to listen which probes beneath surfaces, which hears the interior of speech.

But the ratio of the said to the unsaid extends further than the near proximity of the context and of the depth of the saying. Horizon was first noted as extreme, as limit, and as the Open beyond the present fringe of presence itself. But the further reaches of horizonal significance are not without relation to the proximal horizon. There are occurrences when in word there may be heard an intimation of a wider limit. Such is *poetic word.* Poetic word elicits a new context. It brings to saying what has not yet been said. There is here a sense of violence to word in that the poetic saying disrupts the clarity of the sedimented unsaid.

Poetic word, however, is not merely the novel word. The new word, the creative or poetic word, is not necessarily a word which appears for the first time in the vocabulary of humankind. Perhaps it rarely is. It is rather a word or saying which *opens* experience precisely toward the mystery of the silent horizon as the Open. That which says the horizon is that word which spans the horizon, thus it may be new and old simultaneously. The "linguistic analysis" practiced by Heidegger is often an example of spanning horizons. The methodology which simultaneously "inquires backward" into the very roots of Western thought, into Heraclitus and Parmenides, and which also opens and creates meanings in ancient words which were not at all clear there to begin with is a poetizing thought at the horizon.

The sample of *Dasein* in such analysis is sufficient to suggest the possibility of a wider saying. In its ordinary context, *Dasein* is what is thought of as an ordinary existent or thing. But in Heïdegger's thought *Dasein* becomes *Da-sein* the "*being-here*" which I am. "Being" as an active experiencing and "here" as the finite position which I occupy are my *Da-sein* in a way more significant than the mere "being-there" of an inert object. By opening the word to a wider and deeper context, the word becomes "poetic" in the sense of a bringing-into-being of a meaning; but at the same time it is a bringing-into-being of a meaning which I almost "knew all the time." Philosophical poetizing is such an opening of language-as-word. It is making silence speak. The silence is the horizon, and the word opens toward the horizon. Such is the wider opening which allows significance to be gathered more profoundly.

Is all of this too much to find in the reverberation of the voice? The question of the horizon of silence was posed, as was the question of music, to locate the centrality of word in voice. Ordinary speech, although it potentially contains the richness of the unsaid, in its very ordinariness allows what is hidden to "float" lazily in the midst of its words. Yet even in gossip there lurks the ratio of the said to the unsaid. The possibilities of silence are vast. However, it is in extraordinary voice, the dramaturgical voice, that sounded significance can be amplified. United in a single saying are the "Cartesian" realms of sound and meaning.

CHAPTER FIFTEEN

DRAMATURGICAL VOICE

Voice sounds, but its possibilities are not always amplified clearly. In dramaturgical voice, however, the sounding of voice is amplified. Dramaturgical voice stands between the enchantment of music which can wordlessly draw us into the sound so deeply that the sound overwhelms us and the conversation of ordinary speech which gives way to a trivial transparency which hides its sounded significance. In dramaturgical voice there is united in the same moment the fullness of sound and of significance as a paradigm of embodied word.

The dramaturgical voice amplifies the musical "effect" of speech. This heightens the significance of the word which has been spoken. In the *saying* of dramaturgical voice there is *dabar*, or "*word-event*." This word-event is an occasion of significance which is elevated above the ordinary. A dramatic presence, precisely in the context of being in its set aside, in its elevation over the ordinary, allows what is implicit in all speech to emerge more clearly.

Dramaturgical voice occurs in drama. With the ancient Greeks theater remained quasi religious. Here the voice of the actor emerged from the mask, or *per-sona*. Not only was the individual human face masked or transformed and set aside from the ordinary in a stylized form, but the voice was also masked, transformed, becoming the voice of a god or a dramatic hero.

So, too, does dramaturgical voice occur in ritual or liturgy. The liturgical voice said in the mass, the prayer, and the litany is a transformed voice. The speaker is elevated in minaret, pulpit, and altar; and the extraordinary voice of the cantor or the cardinal speaks in tones which elicit the gods and the saints of the historical community of humans.

Dramaturgical voice also occurs in recited poetry. The celebration of language which emerges in the reading of the poem, the epic, or the legend is another amplification and setting apart from the ordinary which, in turn, renews and enlivens the ordinary. It may be in a voice of protest, as when the voice through poetry asserts the freedom which a regime attempts to still. Or it may be in the simple celebration of a barely noticed aspect of nature, when the poet recites a simple haiku.

In each form dramaturgical voice reveals the possibilities of voice, of voiced language-as-word. Philosophy, however, has often harbored a suspicion of dramaturgical voice. There lies within dramaturgical voice a potential *power* which is also elevated above the ordinary powers of voice. Rhetoric, theater, religion, poetry, have all employed the dramaturgical. The dramaturgical voice persuades, transforms, and arouses humankind in its amplified sonorous significance. Yet from the beginning there is the call to *listen* to the *logos*, and the *logos* is first a discourse. This discourse spans the realm of (auditory) word from the most inner silences of conscience to the uppermost reaches of dramaturgy. Comprehensive listening calls for a listening to the dramatic as well as to the quieter forms of discourse.

But there is another reason to inquire about the dramaturgical voice today. It is the *voice* of dramaturgy which remains at the center of spoken language-as-word. Reason, which at times becomes "voiceless," carries hidden within it a temptation to create a type of disembodiment which becomes a special kind of tyranny forgetful of the human, forgetful of the existential position of humankind. Voiceless, wordless reason becomes the property of an elite whose technical tyranny becomes a threat as great as that of the ancient rhetorician. A reassertion of the role of voice becomes a reassertion of the essential intersubjectivity of humankind as being-in-language.

A phenomenology of the voice is in this sense not only a return to the center of embodied meaning in sound but a return to the existential voice, to the speaking and listening which occurs with humankind. In the voice of embodied significance lies the *what* of the saying, the *who* of the saying, and the *I* to whom something is said and who may also speak in the saying. In the voice is harbored the full richness of human signification.

Thus, not only is there the constant possibility of polyphony in the realm of voiced word, there is also the possibility of a harmonious or disharmonious gestalt in any occurrence of word. Here there is a counterpart within sounded word which reverses the first approximations of sound and sight in relation to the experience of language-as-word. In a panoramic view of the visual field there is an all-at-once quality to the experience. Within the view lies a multiplicity of things united spatially in a gestalt. But in reading, there is a certain "seriality" or "linearity"; writing occurs in a line, and to read is to follow the series. (Speed reading, of course, takes larger gestalts into consideration, but the "linearity" remains in that the order remains important to comprehension.) In listening to voiced word, however, there is a different type of all-at-once gestalt which, although also serial in a strictly temporal sense, is a gestalt in which the harmonics or disharmonics of voice occur. The "meanings" which are more than merely grammatical ones occur within this all-at-onceness giving dramaturgical voice, in particular, its amplified sense of possibility.

To return to a single example, this harmonics of saying revolves around the at least doubled significance which lies in *what* is said and in *how* it is said. The "bare word," *Adam*, here may have as its substitute the simple saying, "I hit the wrong nail!" The amateur carpenter knows all too well the significance of this exclamation, but the voiced word with its bellow of rage contains in its all-at-onceness the sounded significance which exceeds a "bare" exclamation. There is a doubled "grammar" in the sounding, with its "inflections," "intonation," "accent," and "stress," which is the singing of the tongue in its full expressivity. This "grammar" sounded in the how, co-present with the what, of the saying is also part of the voiced tongue.

The obviousness of doubled significance finds heightened awareness in the listening which is amplified, and which is trained to the nuances which ordinary listening does not detect. Thus if the philosopher's listening is particularly acute for the declarative or the argumentative, not all listening is so constituted. There is another listening and speaking which attends to the "grammar" of a different dimension of embodied sound in voice.

The Actor

The actor is such a listener. His ear is selectively tuned precisely to that dimension of voice which utters the *how* in which the saying occurs. His listening as well as his speaking is dramaturgical, and his ear, tuned to those modulations which are already preunderstood among humankind, reflects and amplifies this language of multiple "grammars."

In his apprenticeship he learns to incarnate anger and sorrow, tenderness and pathos. As he learns he notes the multiple ways in which the harsh anger of the shout which electrifies the audience can also give way to the soft whisper of simmering hatred in which the threat of wrath is quite palpable. His voice fills the stage with amplified sounded signification.

His listening is in a sense a purposeful decentering of precisely those things to which the philosopher bends his own ear. The actor's preference for *voicing* is what allows his voice to bring to life the wider context of meaning which animates the drama. Nor is this voicing without its structure. The audience understands in its listening the modulations of the voice which it hears. They are "absorbed by" and "enter into" the sounding words which present the human situation in comedy and tragedy. Here is the embodiment of sounded signification beyond what is merely declarative in which a whole range of unsuspected existential possibilities may come to life.

In dramaturgical voice the transformed and amplified possibilities of sounded significance show forth. The musical qualities of voice are enhanced. This is particularly apparent in the voices of a chorus in which the rhythm, the chantlike repetitions of sounds present a mood as in the chorus of women in T. S. Eliot's "Murder in the Cathedral."

> Here is no continuing city, here is no
> abiding stay.
> Ill the wind, ill the time, uncertain the
> profit, certain the danger.
> O late late late, late is the time, late
> too late, and rotten the year;
> Evil the wind, and bitter the sea, and grey
> the sky, grey grey grey. . . .[1]

The building repetitions of "late late late" and "grey grey grey" paint a sounded background of somberness and anxious foreboding in the voice of the chorus. Against this background the foreground speech of the single character stands out having been given the depth of the overall mood-tone. The chorus is the "continua" of the drama in such instances; the character's voices are the "solo," and drama is the full "music" of voices.

Drama is not only an assemblage of characters with individual voices but a totality which is "symphonic" in its orchestration. Even when there is a "single" character, as in "Krapp's Last Tape," there is a minimal symphonic harmony to the monologue of time and tone in the voice on the tape. *What* is said, the discursive, in voice is never present alone but is amplified within the possibilities of *how* the voice says it.

In this, however, the dramaturgical voice amplifies the possibilities of voice, but its reverberations strike a response which is prepared. The tonalities of voice are foreknown in the listening and the voices of humankind. There may even be a sense in which the tonalities of felt sound are "universal" in a wider sense than the tongue of a particular language-as-word. Crying, moaning, the curse or threat, and certain kinds of singing convey a wider significance which overlaps with the amplification of dramaturgical voice and is foreknown by communities of humans beyond the boundaries of particular tongues.

If what is said is *in* sound which is simultaneously significant in the how of saying, *who* does the saying is also co-present in voice. Here dramaturgical voice reveals a complexity within the *per-sona* of voice. The sounding voice is both a penetration into my self-presence and the presence of *otherness*. But dramaturgical voice raises the question of which other. The actor speaks in a role, and the voice he speaks may in some sense be the voice of another. Oedipus and Hamlet appear, and the actor as an individual recedes. A dialectics of the self and the other appears in the very presence of the drama.

But the self and the other are not on the same plane. The actor's voice does not obliterate the self. There is a style to his voice which remains his own even while the other emerges into the foreground. I still recognize Richard Burton or Lee Marvin or Bob Dylan in the voice of the amplified other who is present as the dramaturgical character. Yet the drama also *transforms* the voice in its amplification: it is not a "mere" Burton or Marvin or Dylan who appears, but the character in the voice of the actor.

In this dialectic of amplified voice lie the possibilities of "bad faith" in roles which Jean-Paul Sartre has often described. The continuum of sameness to otherness which allows us to enact roles belongs also to voice. But the dialectic is not without recognizable structured aspects. His individual style of voice determines to some degree the range of roles an actor may play. One cannot cast a thin, effeminate voice into a male heavy role without some other factors taking effect. But the style of individual voice relates far more to deeper aspects of voiced language. In voice there is something of personal history and situation as well. When I speak I reveal something about my origin, my situation within living language. The Southerner who both hears and says *pin* and *pen* with the same pronunciation lives in speech in a way which is different from those with New England accents.

Nor does the loss or transformation of such accents diminish the significance of one's being in speech. The loss through effort or the acquired loss typical of mobile contemporaries is also significant in that the situation within speech is also changed. The cosmopolitan "general" accent bespeaks a different situation in speech. There is an essential significance to my style of speaking situated in its relation to language-as-word. In Pygmalion fashion, to change one's speech implies a greater change, a change of self. Indeed, the rapid and willful ability to transform voice and its situation within speech takes its own form among the community of actors. To "become" another in voice with ease opens the way to a certain self-becoming which may, and often does, emerge either as a self-doubt reflexive toward one's self, or an ease which masks or, better, allows for the making of an extraordinary flexible character to which others fail to find ordinary stability adhering.

Yet every voice is also trans-individual. The style which adheres to a role is transcended in the role in which otherness may appear. The adaption of an accent not mine, the transformation from twentieth-century America to sixteenth-century England is a possibility of dramaturgical voice presence. Yet a personal history and style is not discontinuous with the trans-individual in voiced word. Language-as-word is not private or individual as such but is intersubjective from the outset. This is also revealed in speech, for in speaking I always show more in saying than myself alone.

Dramaturgical voice plays within the intersubjective possibilities of language. Otherness shows itself in the roles and "voices" of Falstaff and Everyman, but such possibilities are inherent in the voices of language from the beginning. This multiplicity is threatening to a concern with an "authentic" voice which at base is a concern with a *single* voice. The demand that the innermost voice be the same as the outermost voice, that only one role ever be played, harbors a secret metaphysical desire for eternity and timelessness.

Dramaturgical voice does not display the difference between appearance and reality so much as it does the multiple possibilities of every voice transformed from ordinary to extraordinary. The "others" who appear are the human possibilities which are also "my" possibilities, and the drama is a "universal" play of the existential possibilities of humankind. But the drama is the extraordinary in the sense that in the set-aside time of the stage the existential-imaginative possibilities portrayed there are not bounded by the single life I have to live. The drama is an exercise in imaginative variations which portrays the range of possibilities open to humanity but not to a single individual who is temporally bound and limited. The elevation of humanity in its full complexity thus makes us aware of the *tristesse* of finitude. The cartharsis is also the recognition that the play does *not* go on forever.

Dramaturgical voice also amplifies the previously noted phenomenon of the auditory *aura* of the presence of the other. The actor amplifies the sounding voice, he *projects* voice into the recesses of the theater. This resonant voice is an auditory aura which im-presses in sound. The auditor is not merely metaphorically im-pressed, but in the perception of the other in voice he experiences the embodiment of the other as one who fills the auditorium with his presence.

Yet what is most dramatic in the projecting, resonant voice from the stage is also present in the experience of the other in ordinary discourse. The other is more than an outline-body. In speech and the experience of voice there adheres an enriched experience of the other. The person with a strong voice is impressive in a way that the person with a weak voice is not. Contrast the stage presence of the accomplished actor with dramaturgical vocal power against that of the small child in a school pageant. The faltering lines, the uncertain quaver, the lack of resonance and projection bespeak the dependency and childlikeness of the actor. Much of the charm of the pageant, although best appreciated by the parents, lies precisely in the pathos of the not yet fully matured voices of the small children.

In the voice of the actor, the drama of the other comes to presence. But the other is bound to me as well. We both stand and take our speech *within* sounding language. Sounded language surrounds and penetrates the recesses of the self and the other. The dramaturgical voice amplifies and displays these variations upon the modes of being in language.

The Liturgist

What I have written of the dramaturgical voice of the actor may also be applied to other dramaturgical forms. But if the actor learns to personify the other, particularly the other human, the liturgist emphasizes another possibility of voice. He bespeaks tradition and the voices of the gods. They sound in his voice. In the West, with its religion of Word, the sound of dramaturgical voice is particularly marked. The marks which have been added to the Torah, to the written form of the tongue which has vowels in speech but none in writing, reflect the ancient tradition of the synagogue in which the Word must be correctly canted. "Hear, O Israel" is voiced in the event of the liturgy and in a traditional and stylized sound.

The liturgical voice also reads the Gospel, setting it apart from ordinary discourse by the saying, "Thus beginneth the lesson." The moment of the elevation of the host is also the moment of setting apart in the sound of chime or bell. Even the sects which claim to have dispensed with liturgy retain a stylized dramaturgy of voice. The repetitious but dramatic voice of the Pentacostal preacher sounds a call for repentence similar to the muezzin's older stylized call to prayer in Islam.

Liturgical voice is in certain ways less flexible than the dramaturgical voice, but that is because its transformation of voice echoes a more distant type of saying, a highly traditionalized otherness. In liturgical voice there are echoes of the ancient past. Matthew, Mark, Luke and John; Abraham, Isaac, and Moses speak in the liturgy. But the ultimate otherness of liturgical voice is the echo of the gods.

There is particularly in Western religions a theology of *Word*, where God himself is the *God of Word*. God *speaks* and the world comes into being. In Christianity Jesus *is* the *Logos*, or Word. The Holy Spirit in *filling* the congregation creates a unity of spirit, a unity of the *tongue*. The presence of Word is central to Western theology.

More subtly, the very experience of the God of the biblical traditions is an experience of word in voice such that the person of God is "like" an intense auditory experience. The Western God of Word gradually became known as an omnipresent but *invisible* God. In the ancient Hebrew traditions it became totally prohibited to *represent* God in an image. To "image" God was to create an idol; but while saying the name of YHWH was also prohibited, there was no prohibition of the "speech" of God being "represented" in the written word. God's Word must be remembered, engraved upon the heart. The invisible God was not absent, but *present* in Word. When he was absent it was when he *refused to speak*.

But omnipresent, invisible presence is presence of sound in its most dramatic moments. The liturgist fills the synagogue with sounding voice. And in the classical religious experience of Isaiah in the Temple, vision is obscured as the temple is filled with the smoke of the offering, but the voice of God presents itself in the very midst of the visual obscurity.[2] The God of voice surrounds, penetrates, and fills the worshipper.

The God of voice, of Word, is also "personal," for the voice bespeaks a per-sona. The anthropomorphic quality of the arguments between God and Abraham, the covenant between God and his Chosen People bespeak the incessant discourse between men and gods. In a theology of Word both men and gods belong to a conversation.

There is also in the theology of Word a distinctly temporal-historical dimension. Word is *dabar*, which is both "word" and "event." God's speaking is an event which is itself an act. The willfulness of the

Western God is the temporality of his speech-act, the manifestation of the God of historical moments. The making of those words "come to stand" in the Torah and the Bible is the temporal-historical equivalent of stability, of "essence," but with a difference. Word is to be repeated, to be remembered, to be reenacted in the liturgical event. The Feast of Tabernacles, the Last Supper, the days of the liturgical calendar are at base reaffirmations of the historical.

The essentially invisible presence, the surrounding and penetrating presence, the temporal-historical presence of Word, of holy voice is also a *dominantly auditory presence* within the heart of Western theology. This God cannot be coerced but must be "let be," for he speaks or "shows" himself in Word only when he will, just as sound occurs when it will. Such a God is dramaturgical voice in the extreme even if the voice at times must be heard as the "still small voice" which sounds silently after the thunder and the hurricane.

In the drama of the liturgy the god is experienced *in* the presence of voice. But the listening which no longer hears *in* voice the sounding of the god cannot at will draw speech from the silence. The god is the ultimate extreme of otherness which nevertheless belongs to the same possibilities as sameness in the presence of word.

The Poet

In its most ancient form, poetry was spoken, recited in a singing of verse. There was the recitation of an epic, the singing of a ballad, the extended storytelling in the form of verse. The music and rhythm of poetry retains its adherence to the spoken word, for poetry is "close" to music as a form of dramaturgical voice. In one respect, the poetic voice is the most flexible of voices. Its range extends to the liturgical in the psalm and the theatrical in epic. But more than giving presence to the voices of others and of gods, poetic voice extends to things. In poetic voice there resounds a "speech" of rocks, mountains, and sky; of machines and jugs and other voices of the world.

Even in written form poetry retains its adherence to the sensuousness of sound. It is the sounded significance which sings in the mystery of a beast portrayed by Blake:

> Tyger! Tyger! burning bright
> In the Forests of the night,
> What immortal hand or eye
> Could frame thy fearful symeetry?[3]

And although sound is more explicitly mentioned in Frost's "Sound of the Trees," its adherence to the word is one which elicits the sounding of trees.

> I wonder about the trees.
> Why do we wish to hear
> Forever the noise of these
> More than another noise
> So close to our dwelling place?
> We suffer them by the day
> Till we lose all measure of pace,
> And fixity in our joys,
> And acquire a listening air. . . .[4]

But although not lacking in any form of well developed dramaturgical voice, the voices of poetry perhaps make more apparent a certain sparseness of speech which reveals another side to being in language. Poetry amplifies silence. In a directly elicited sense of silence in poetic utterance there is the haiku:

> I heard the bird
> in the valley
> and suddenly
> realized the silence

Here, the sensuous richness of experience is elicited in few words. I "hear" in the poem the call of jay or crow over the mountains, and in the call and its echoes I realize the surrounding silence which allows the call to "stand out." The call is indeed more stark and vivid in such sparseness of expression.

This suggestive simplicity may, of course, be noted in specific forms of voice. The radio drama recalls nostalgically for many a sense of richness of imagination often lacking in more explicit audiovisual presentations. A contemporary illustration is the Stoppard "Artist Descending a Staircase" in which sound alone conjures the context, suggesting and making imaginatively present the absent global quality of experience. In more traditionally stylized form the very meager or even silent quality of the No drama of Japan elicits the richness of silence which is not empty but filled possibility.

The descriptive and enabling power of the poetic is in each of these cases "richer" than many forms of direct or extended analytic description. In the poetic voice a "gestalt" occurs which engages the hidden horizonal significance of that which is present in the speaking. The poetic voice "allows" the horizon to be "given," to "e-vent" itself in and around the words. Here I write of the poetic *experience* prior to any possible analysis of the "mechanisms" or "techniques" which a metaphysics of poetry seeks to discover.

The word of poetry enables otherness to be vividly present, even that hidden significance which emerges in horizonal significations. The elicited and suggested "noema" of that which appears is of course strictly correlated with the "noesis" of the listening act. The poem calls for its own form of listening which is, as in every listening, enriched by both the wealth of the listener's own experience and the ease with which the poem may be "let be" in its significant presence.

Dramaturgical voice amplifies the possibilities of sounded significance throughout the full range of human voice. The gods, things, and others gain voice, and all are situated within the silence which is the horizonal limit of sound. This extended amplification displays the conversation which is humankind.

1. T. S. Eliot, "Murder in the Cathedral" (London: Faber and Faber, 1971), p. 18.
2. Isa. 6:1-10.
3. William Blake, "The Tyger."
4. *Robert Frost* "The Sound of the Trees."

CHAPTER SIXTEEN

THE FACE, VOICE, AND SILENCE

In the very midst of the conversation which is humankind there are beginnings. But not all beginnings belong to the center of language-as-word. There are beginnings which occur before and after speech. If I meet an other who is a stranger who may not speak my tongue, then the meeting is one which takes place within "language" only in the broader sense of language as significance. Here, decentered from the focal, clear meaning of word, I meet the other as *face*. But the face, too, belongs in its own way to the *play of the polyphonic* which exists at the heart of voiced language-as-word. To meet the other as face includes the possibility of conversation. The face in its signification bespeaks, in relation to the center, the pregnant nearness of significant silence.

From attending to the heightened amplifications of dramaturgical voice, I return to the ordinary affairs of daily life. Perhaps I am engaged deeply in some absorbing project, and I fail to hear the other enter the room. I suddenly see him and look at him face to face. He has broken the solitude, and on such an abrupt occasion there may indeed be something like the "internal hemorrhage" which Sartre describes when my "world" bleeds away under the gaze of the other.[1] In the face-to-face meeting, however, this "hemorrhage" is brief, and the shock generated by the other gives way to an *invitation to word*. This is so even if the word is perfunctory. Face to face meeting without any word results in awkward silence, because in the meeting there is issued a call to speak.

What follows, of course, is variable in relation to the degree of mutuality between us. If the other is a salesman who intrudes not only bodily, but with mock intimacy in his voice and too soon uses my familiar name and too soon "violently" presumes mutuality, I may retreat or reject his call. If, on the other hand, the other is my friend, the conversation begins by already presupposing the long history of what has been said previously, and the brief ritual greeting gives way to discourse which moves easily and freely. And if the other is my beloved, then the conversation which has already occurred over a long period is deepened by the richness of the unsaid, and a sparse economy of words conveys more than another party could hear. The pregnant silence of the unsaid allows the horizon of significance to carry the burden of the conversation in its greater ratio to the saying.

In each of these cases the presence of the other as face carries both the significance of pregnant silence and of a call to speech and listening. The silent call of the face may give way to spoken word which then presents itself with all the surrounding, penetrating power of sound in a call which insists that I "obey" by responding. The ancient meaning of *obaudire*, "to listen from below or from the depths," echoes in the call. The other exceeds the silent presence of the face in the aura which has been cast and which places both of us in the midst of mutually penetrating sound.

This meeting which gives way to conversation however, is not simple but primordially complex. It takes place not only in the "dialectics" of the face and the voice, but it echoes the *play of polyphony* which sets the limits for existential possibilities. Only ideally and rarely do I attend so solely to the other that the full mutuality of being-in the fullness of language may occur. Equally, I may retreat into myself and the self-presence of my inner speech whose "static" closes off some of the call of the other. Every conversation can not only mask itself in the ambiguities of word, in the ratio of the said to the unsaid, but it can flit among the possibilities of the polyphony of voices we are. Thus the meeting is fragile.

There may be a sense in which philosophy, ever seeking the stable and "eternally" secure, detects and resents this fragility. Perhaps it is even implicitly aware of the polyphony. Philosophy has often resisted recognizing polyphony as primary. Philosophy's desire and aim has been for a *single voice*, identical "within" and "without," which harbors no hidden side of unsaying or of countersaying.[2] In its visual metaphors its goal has been a pure light or transparency; auditorily its goal is a sound which does not harbor a relation to the silences which conceal a hidden dimension to every sound. But the single "authentic" voice occurs only in certain privileged moments. Those are the moments of fragile meeting in which there is an exchange of concentrated listening and speaking. There remains, though, an important point in relation to the primordiality of polyphony which sometimes escapes the philosophical traditions which have also harbored deep suspicions concerning voice. The very possibility of an essentially "doubled" voice is a possibility which holds that every "expression" also hides something which remains hidden and thus cannot be made "pure."

This remains the case even for the ideal moments of genuine speaking and listening. The speech situation occurs within the context of full significance. Here not only voice but the face as the indicator of pregnant silence remains part of the entire gestalt. This is the *nearest* horizonal aspect which surrounds the central speech. But there also remains the hiddenness of the "silent" voice of inner speech which, like the hidden side of a transcendent thing, remains hidden to the other. And beyond both the pregnant silence bespoken by the face and the "outer" silence which does not reveal inner speech, there lies the *Open* silence of the ultimate horizon. In all three of these respects there remains a hiddenness which belongs to the center of voiced language. The perfectly "transparent" eludes the desire of philosophy.

To the seeker for "transparency" the hiddenness appears as a weakness, a barrier within language; and the dream of overcoming polyphony takes the form of overcoming the Tower of Babel. But this dream itself harbors its own type of darkness, not knowing that the Tower is itself more expressive of the human situation than its hidden presumption of innocence which lies on the other side of existence. The dark desire of the dream emerges from time to time in the very strategies which seek to make absolute the totally transparent. Sometimes these take the form of various reductionist attacks upon language.

The attempt to clarify, noble in itself, can contain a desire to control. This control, however, is at the same time an attempt to overcome the finitude of the play of presence-hiding which occurs in voiced language-as-word. To combat this play various "therapies" are invented which seek to reform, to remake, to transform language in a direction which lies distant to the latent richness of existential polyphony. The symptoms may be seen in the science-fiction nightmares of rationality gone mad. These take the form of *forcing* polyphony into monophonic single speech. The ultimate control of language would, in a sense, be a powerful weapon in controlling thought were it not in part due to the constant possibility of its escape within polyphony. The fantasies of spy stories in which chemically induced "truth" pries into inner speech, forcing it into the open, is but a rape fantasy directed at language-as-word. Closer to frightful reality, though, are the emergent political-behavioral sciences which, through polygraph, blood pressure, and other physical measurements, seek to detect the deeper response, an "inner" response, from those willing to submit to the poll.

In spite of the genuine insidiousness of such a "science" which lies in its ability to persuade persons to submit for the sake of "science" to such a control, so long as there remains polyphony there remains the possibility of a *refusal.* The call of the other is a "command" but a command which may be rejected by the possibilities of silence.

The refusal may itself be neither obvious nor dramatic. It may lurk in the trivial response, in speech which is empty and inane. The implicit recognition of this provides a stimulus for a more coercive demand on the part of the questioner. Thus the refusal may make its appearance more dramatically as the stony silence which elicits a horizonal possibility.

What is more "closed" than the silent refusal? The prisoner brought before the dictator's police sits in silence, his voice does not reveal the secrets which he holds within himself. Yet the silence also "speaks" as the horizonal pregnant signification of the face: it affirms the silence of the interior, one aspect of the hidden which is essential to all speech. Silence lies "close" to speech in such a refusal. The dictator's police turn to torture to extract what is hidden, but the forced response, the forced confession remains inauthentic even if sometimes it attains the dictator's objective. The refusal lies at the horizons of speech; it is where language-as-word retreats to the interior but also where silence itself reveals the significance of the horizon.

If there is an ethics of listening, then respect for silence must play a part in that ethics. But ethics is even more fragile than the ideal moments when a conversation becomes the exchange of single voices and open listening, because ethics must depend upon a silent agreement of humankind which itself already presupposes a certain unanimity of voice and thought. Ethics must take its place second to a sense of community. Nevertheless, it remains that respect for silence is essential in spite of the fragility of such an ethic.

Yet in the very midst of the fragility of ideal listening and speaking, in spite of the double fragility of an ethics of listening, and despite the continued and predictable continuation of the reductionist strategies which seek to control language-as-word, there remains what is easily forgotten. The "darkness" hidden in voiced language is in fact not a weakness of word but its strength. It is the ultimate withdrawing Openness of the silent horizon as full ontological possibility. The relation of voice to inner speech and to the pregnant silence of the face gives way ultimately to the Open horizon of silence. Here is constituted in effect an *ontology of listening and voice* in the sense that there is a permanent set of existential possibilities which exceed the strategies seeking to control or deny them.

The richness of language-as-word thus lies not only in the focal clarity which can be developed despite its own "darker" side which forgets the Openness that situates whatever clarity is attained, but also in the untold and unsaid possibilities of that silence. In this, language-as-word is a kind of music. When a composer or creator of music begins to combine the sounds which will be the musical "statement," he may, as it were, "out of nothing" add sound to sound. The Openness of silence allows this even to the extent of the creation of a cacophonous music. The "final" limit is not reached except in silence. A noisy music which became so cacophonous and intense, to which "nothing more" could be added, appears itself as a kind of silence

The more restrained music which in turn allows pregnant silence to situate the sound nevertheless retains its own relation to the horizon. And in the relation there remains the structure of presence to the surrounding, withdrawing, but Open horizon which always allows a further possibility.

It is in this respect that the poet, like the musician, will always have a *further word*. The creation of a new opening remains a relation to the Open. In this there is no necessity to create a new word as such any more than there is a necessity for the musician-composer to create a new set of instruments, although both such possibilities may be grasped. It is rather a matter of discovering a different gathering of words which allows a new possibility, a new relation to the Open, to emerge.

A further word, compared to the desire for completion and closedness, always remains "beyond" the central clarities which conceal their own relations to silence, but a further word also remains penultimate, because *there is no last word*. The last word, rather, is no word as such. It is the withdrawing Openness which is the "other side" of word but which is bound to every word. The deepest and most profound listening hears not only the voices of the World, it is a waiting which is also open to the possibilities of silence.

Within the "music" of language-as-word, the penultimate word of the poet which is matched by the penultimate musical "statement" of the composer remain but other cases of extreme examples, of amplifications of the existential possibilities of daily life. Every conversation, every meeting of the other hides within itself the possibility of a beginning. This beginning may be as prosaic as the generation of new sentences which the linguist today recognizes as a problem in the understanding of speech. Or it may be more intimate as in the beginning of a conversation which opens a friendship for a longer conversation. But as the beginning it is a beginning in the midst. Beginnings occur within the whole range of language. When they occur in the midst of language-as-word there remains the hidden pointer to the forms of silence, the pregnant silence bespoken by the face, the "outer" silence which masks inner speech, and the ultimate horizon of silence as the Open. In this sense the beginning of man is in the midst of word, but word lies in the midst of silence.

1. Jean Paul Sartre, *Being and Nothingness*, trans. Hazel Barnes (New York: Philosophical Library, 1956), pp. 252-73.
2. For a visual counterpart, see Samuel Todas, "Shadows in Knowledge," in *Dialogues in Phenomenology* (The Hague: Martinus Nijhoff, 1975), pp. 86-116.

AUTHOR INDEX

PERMISSIONS

Permission granted by Penguin Books Ltd. for use of "Tyger! Tyger!" from William Blake; *A Selection of Poems and Letters*, edited by J. Bronowski, 1972.

Permission granted by Holt, Rinehart and Winston, Inc. for the use of "The Sound of Trees from *The Poetry of Robert Frost*, edited by Edward Connery Lathem, 1969.